Lasne R.

D

Jeanloo

le Cullot

C

E

D

la Haye

Frischermont

Couture

Smouhen

E

F

B

G

F

Aywiers

G

F

G

G

Planchenoit

REFERENCE.

A.A.A. *British Position on the Morning of the 18*[th]

B.B.B. *French D*[o]. .... *D*[o].

C.C.C. *March of Prince Blucher thro' Ohain to form a junction with the British.*

D.D.D. *Advance of Gen*[l] *Bulow's Corps from S*[t] *Lambert to occupy their Covered Position*

E.E.E. *Advance and Charge of Gen*[l] *Ziethen's Corps towards the Close of the Battle.*

F.F.F. *Attack of the Prussians on the Enemy's Right Flan*[k]

G.G.G. *Movement of the Enemy's Reserves to oppose the Prussia*[ns]

MAPS OF WAR

MAPS OF WAR
Mapping Conflict through the Centuries

JEREMY BLACK

C
CONWAY
BLOOMSBURY
LONDON · OXFORD · NEW YORK · NEW DELHI · SYDNEY

Conway
An imprint of Bloomsbury Publishing Plc

50 Bedford Square 1385 Broadway
London New York
WC1B 3DP NY 10018
UK USA

www.bloomsbury.com

CONWAY and the 'C' logo are trademarks of Bloomsbury Publishing Plc

First published 2016

© Jeremy Black, 2016

British Library Cataloguing-in-Publication Data
A catalogue record for this book is available from the British Library.

Library of Congress Cataloguing-in-Publication data has been applied for.

ISBN: HB: 978-1-8448-6344-0
ePDF: 978-1-8448-6462-1
ePub: 978-1-8448-6463-8

2 4 6 8 10 9 7 5 3 1

Designed by Nimbus Design

Printed and bound in Malaysia by Tien Wah Press (Pte) Limited

Bloomsbury Publishing Plc makes every effort to ensure that the papers used in the manufacture
of our books are natural, recyclable products made from wood grown in well-managed forests. Our
manufacturing processes conform to the environmental regulations of the country of origin.

To find out more about our authors and books visit www.bloomsbury.com.
Here you will find extracts, author interviews, details of forthcoming events
and the option to sign up for our newsletters.

For Michael Joy
First among neighbours

Author's acknowledgments

I am most grateful to my editor, Lisa Thomas, who is as effective and efficient as she is calming and helpful. We worked together on the *DK Atlas of World History* and I am delighted to see that the team still works well. Thanks too to Sonja Brodie for copy-editing the book, to Jen Veall for her creative picture research and Nicola Liddiard for her elegant design. I have benefited from the comments of Heiko Henning on an earlier draft and am also grateful to Yingeong Dai, Mike Dobson, Spencer Mawby, David Parrott, Karsten Petersen, Kaushik Roy, Don Stoker, Stag Svenningsen, Ken Swope and Harold Tanner for answering questions on particular points.

Contents

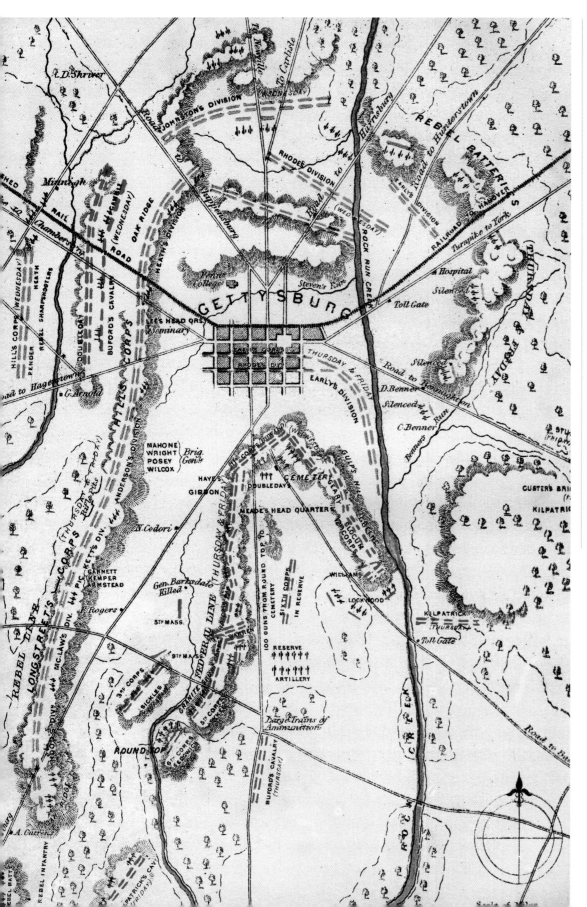

MAP OF THE BATTLE OF GETTYSBURG, 1–3 JULY 1863   An engraving from volume III of *The War with the South: a History of the Late Rebellion*, by Robert Tomes, Benjamin G. Smith, New York, Virtue & Yorston, three volumes, 1862–1867. **LEFT**

ORDNANCE SURVEY MAP OF UTAH BEACH, JUNE 1944
The map was annotated by US General Raymond Barton, commander of the 4th Infantry Division, during the D-Day invasion of Normandy. **TITLE PAGE**

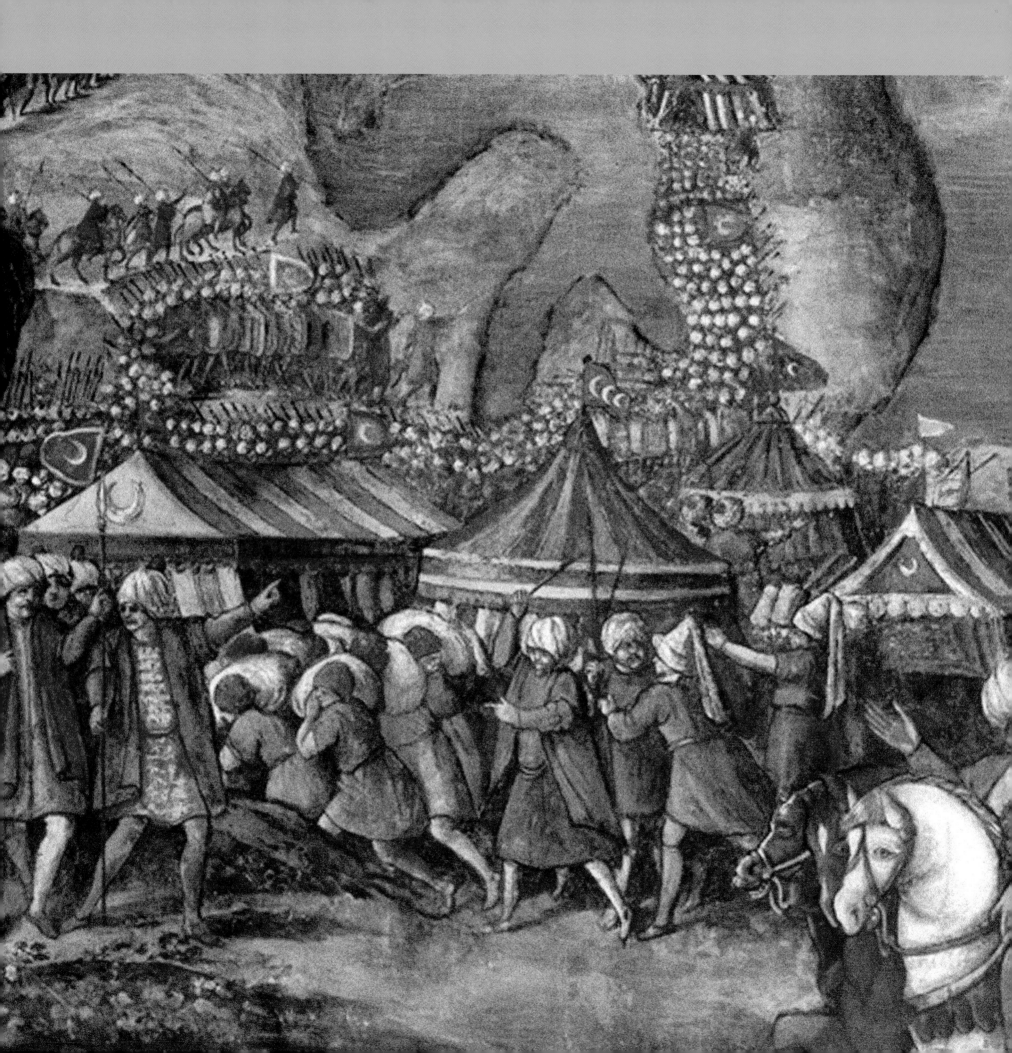

FIRST NOTIONS *of* MILITARY MAPPING

WAR REQUIRES A SPATIAL as well as temporal sense and awareness. At the tactical level, that of battles, it requires the detailed understanding of the relationships between place and terrain, notably height, slopes and water features such as rivers and marshy ground. Each of these was, and is, significant for the tasks set for military units and for their relative effectiveness in discharging them. At the operational level, that of campaigns, the conduct of war requires an understanding of routes and of the relationships between place and terrain over the area of campaign. These are necessary for effective planning both about advances and concerning supplies. At the strategic level, there are requirements for geopolitical information so that challenges and opportunities can be grasped and priorities determined.

All of these levels involve the equations of force and space, and each requires maps: maps to present, maps to understand, and maps to convince. This book covers the variety of different ways in which maps have accompanied war, notably for geopolitical consideration, strategic planning, operational purposes, tactical grasp, news reporting and propaganda. In doing so, it considers the different stages of conflict – preparation, planning, initiation, waging, outcome and retrospect – and assesses the manner in which maps framed a particular perception of war at various scales and to very different audiences, from strategists to tacticians, from the military to the public, from those who were present to those who were distant. The application of mapping technologies developed in peacetime is assessed, as is their use in wartime. The entire situation in recent centuries is one of dynamism, as opportunities are grasped, tested and used.

For most of history, and still to this day, the relevant maps have overwhelmingly been mental maps, the understanding of place in the mind's eye. These were

ready means to help consideration and exposition. For example, the plan of an ambush or a fort might be drawn with a stick, a finger in the dirt or in powdered sand, sketched in the air or described with words. These methods did not leave records, written or otherwise, that survive, and that is a crucial point when looking at the mapping of war. It is a particularly significant point for pre-modern warfare. Nevertheless, pre-modern armed forces carried out complex operations that would have required a foreknowledge of terrain. How that knowledge was conveyed is less clear, but oral report was the key

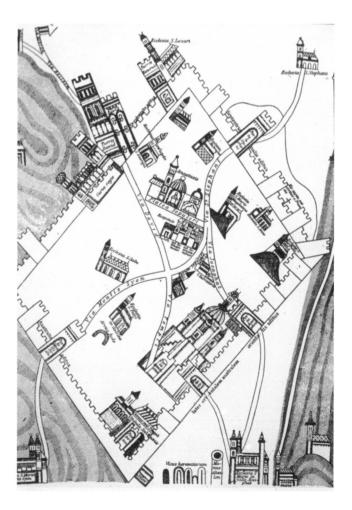

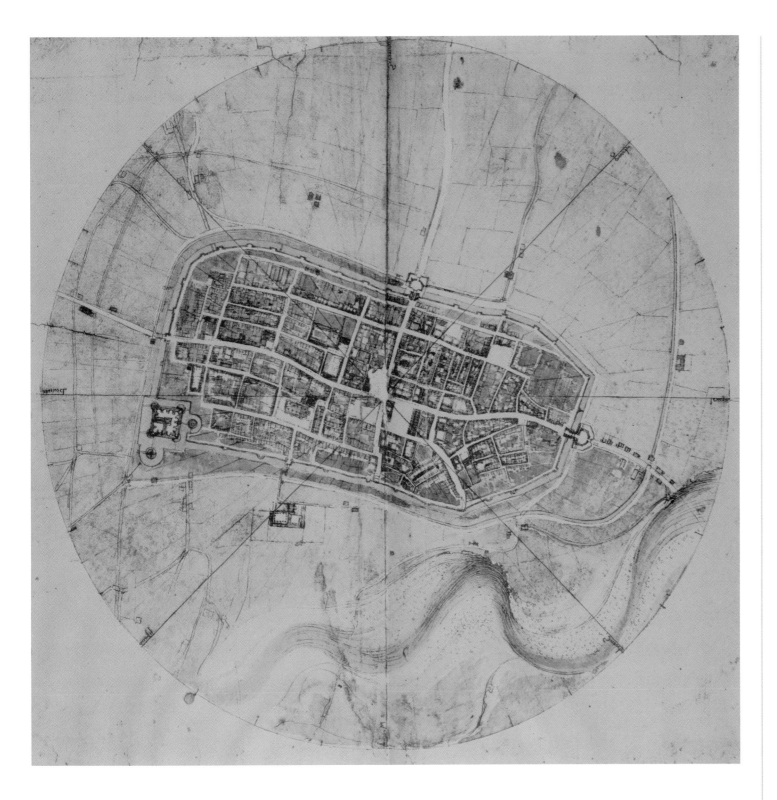

DRAWING OF IMOLA. BY LEONARDO DA VINCI, 1502 The strength of fortifications had become a more prominent issue once the Italian Wars broke out in 1494, with cannon playing a role in the successful French invasion of Naples that year. This began a period of conflict in which Pope Alexander VI, eager to expand the Papal States, instructed his second son, Cesare Borgia, to subjugate the region of the Romagna in eastern Italy. Cesare Borgia started by capturing the town of Imola in 1499 and continued to make conquests until 1503, when the death of his father was followed by his arrest under the orders of the new Pope, Julius II. Da Vinci was employed to survey Imola and to suggest how best to improve its defences. In Da Vinci's aerial view, the city emerged clearly as a defensive system. The significance of the moat was readily apparent. This approach was much more effective than that of adopting an oblique perspective and displaying buildings in elevation. LEFT

THE OTTOMANS VERSUS CHARLES V, 1535.

ENGRAVING BY FRANZ HOGENBERG In 1534, Tunis was captured from Mulay Hasan, its pro-Spanish ruler, by Hayreddin 'Barbarossa', the Ottoman Grand Admiral and the key figure in Algiers. This was part of the Ottoman attempt to consolidate their power in North Africa after their conquest of Egypt in 1517. The Emperor Charles V, who also ruled nearby Sicily as well as Spain, responded in 1535 with a major expedition that included 82 war galleys and more than 30,000 troops. The expedition, which reflected Spanish capability for force projection, was in large part paid for with Inca gold from South America which repaid loans from Genoese bankers. Mounted in ferociously hot conditions, this expedition displayed amphibious capability and success in fighting on land. The fortress of La Goletta at the entrance to the Bay of Tunis, although defended by a large Ottoman garrison, was successfully besieged, falling on 14 July. A week later, Tunis was captured and sacked. Thousands of Christians were released from slavery, while thousands of the local population were slaughtered and large numbers sold as slaves. Charles installed a pro-Spanish Muslim ruler. This was a high-water mark for Spanish attempts to bridge the western Mediterranean, which did not appear to contemporaries as a border. RIGHT

means. In the Roman sources, there are not references to the use of maps on campaign, but there are many references to scouts being sent out to learn about the locality. This suggests that an aural and visual approach to 'mapping' was used, rather than a written system.

Oral report went on being significant, but sometimes with disastrous effect, as during the Crimean War when, as a result of misunderstanding orders, the British Light Brigade cavalry charged directly into Russian artillery at Balaclava in 1854. This was a classic instance of poor 'situational awareness' and was not redeemed by the bravery displayed.

MAPS IN THE CLASSICAL WORLD

The degree of spatial depiction at the strategic level is far from clear. References to maps that were not for specific military purposes, notably maps of the known world, do survive from early civilisations and, in part, they reflect the results of wars. In particular, the spread of territorial control through conquest, a process that was the major cause and consequence of war, ensured that far more geographical material became available. The major extensions of the world readily known to Classical commentators (thanks to the conquests of Alexander the Great and, later, the Romans) provided

SIEGE OF MALTA, 1565. ITALIAN SCHOOL, ENGRAVING

An epic much depicted in Christian Europe, this campaign was difficult to reproduce in one image as it lasted for many years. Süleyman the Magnificent sent a powerful expedition of 140 galleys and about 30,000 troops to capture Malta, the principal Christian privateering base in the Mediterranean and a threat to Ottoman trade. The defenders, under Jean de la Valette, the Grand Master of the Order of St John, had only 2,500 trained soldiers, but also local levies. Landing on 18 May, the Ottoman attack failed because of the courage and tenacity of the defenders and because of the failure of the Ottoman land and sea commanders to agree and implement a coordinated and effective command structure and plan. The summer heat was also a factor, as were logistical difficulties, including the supply of drinking water. Although the Ottomans captured their initial goal, St Elmo, one of the forts at Valletta, the garrison fought to the death but the others, Senglea and Birgu, held out. The heroism of the defenders of St Elmo, which delayed the Ottomans for two weeks, was crucial to the logistical problems the attackers encountered and gave the Spaniards more time to prepare relief attempts. A Spanish relief force, which landed on 7 September, tipped the scales after the Ottomans had failed to crush the defence. Unwilling to face the new foe, the Ottomans, having lost perhaps 24,000 men (against 5,000 defenders) retreated. They never again attacked Malta. LEFT

geographers such as Eratosthenes and Strabo with much fresh material and ideas, the two frequently being linked.

Roman territorial expansion, and the accompanying and subsequent need to protect the frontiers of Roman rule, helped lead to a major increase in geographical information for the Classical world as a whole, as well as to improvements in the accuracy of maps. Power was also served, as maps provided a way to display strength and purpose and to record success. Prestige was always a key element in mapping, as in war. This presumably was why Julius Caesar was (later) held to

have ordered the surveying of the known world. Military maps were probably among the sources used by the important geographer Ptolemy (c. 90–c. 168 CE), but these maps have not survived.

Spatial information was probably produced in part in lists. Vegetius, the author of the fourth-century CE *Epitoma rei militaris* (a summary of the art of war that was also influential in the European Renaissance of the fifteenth century), stated that a general must have 'tables drawn up exactly which show not only the distances in numbers of steps, but also the quality of the paths, shorter routes, what lodging is to be found

PICTORIAL MAP OF THE 1529 TURKISH CAMPAIGN.

BY JOHANN HASELBERG AND CHRISTIAN ZELL, 1530

This map of the Ottoman advance into the Balkans captures a sense of menace. In 1529, the Sultan, Süleyman the Magnificent (r. 1520–66), who had crushed the Hungarian army at Mohács in 1526, advanced on Vienna, taking Buda en route. However, Vienna proved the limit of the Ottoman range. Süleyman did not reach the city until 27 September, and a determined defence was able to resist assaults until the Ottoman retreat began on 14 October. Campaigning at such a distance from their base caused major logistical problems, as troops and supplies had to move for months before they could reach the sphere of operations, and the onset of winter limited the campaigning season. During the campaign, Ottoman raiders reached as far as Regensburg (in modern Germany) and Brno (in Czech Republic). Vienna was not besieged again by the Ottomans until 1683, but, unlike then, Ottoman failure in 1529 was not due to defeat by a Christian relief force. In the bottom left, Emperor Charles V, Charles I of Spain, is the leader of much of Christendom. Published in 1530, the map was a call to action. It was accompanied by a booklet pressing for a crusade. **LEFT**

there, and the mountains and rivers'. The most frequently reproduced of Roman maps, the *Tabula Peutingeriana*, a twelfth-century copy of a map made between 335 and 366 CE, showed relative locations rather than exact positions. It recorded main roads as well as other features, including mountains, rivers, forests and staging posts. Such maps would have been useful for planning the movement of troops, a classic instance of fit-for-purpose mapping. More generally, a high proportion of early maps was in the form of linear itineraries and, therefore, lacked the measured space and coordinate geometrics we now associate with mapping and maps.

EARLY ASIAN MAPPING

The situation was similar in China, where information obtained from Chinese overseas expeditions and from border wars fed into a long-established practice of assembling such material. This culture involved maps as well as lists. Under the Han Dynasty (206 BCE–220 CE), there was an official dedicated to surveying in preparation for war. Chinese colonisation of steppe lands to the north and north-west, a colonisation closely linked to defence and forward-defence against nomadic attacks, relied heavily on an understanding of topography and invasion routes. During the Sung period (960–1279 CE), the Northern Sung empire accumulated, mapped and stored geographical information on the empire.

From the late eleventh century, however, the relevant government apparatus declined and mapping was largely taken over by private scholars. Printed maps of the empire circulated in atlases in the twelfth and thirteenth centuries. These maps encouraged a sense of loss in the face of 'barbarian' advances, an image made use of by poets. It is unclear how far maps played a major or detailed role in Chinese military

BIRD'S-EYE VIEW OF VIENNA DURING THE OTTOMAN SIEGE OF 1529. BY HANS SEBALD BEHAM, **1538** This circular map by Hans Sebald Beham was published as a woodcut print by Niklas Meldemann in Nürnberg. The appearance of this map indicated the widespread interest in the siege. The map was sketched from the spire of St Stephan's cathedral, the circular design emanating from that central point. The city's 300-year-old walls are flattened out in the view so that the inner surfaces are visible, thus enabling defensive armaments to be depicted. Niklas von Salm, the defending commander, conducted a vigorous defence, which contrasted with the weak defence of Belgrade in 1521 by a small garrison that rapidly capitulated. Salm blocked the city's gates, reinforced the walls with earthern bastions and an inner rampart, and levelled any buildings where it was felt to be necessary for the strength of the defence. The 1529 campaign failed to intimidate the Habsburgs who, instead, advanced to besiege Ottoman-held Buda in 1530. LEFT

planning, but it is likely that they did.

In Korea, map-making was not well developed, but there were maps as well as drawings of battles. A Japanese soldier who fought in Korea in the 1590s drew a battle plan for the siege of Namwon, which was taken by storm in 1597 after four days of siege. The drawing is schematic.

In Oriental cultures, maps formed an aspect of the understanding and use of space, both of which had a spiritual character in the shape of geomancy or *feng shui*. The specific positioning of fortifications, as with other buildings, was important to their effectiveness. For example, fortifications could take the place of missing hills to produce a geomantic pattern that was effective in defence. The enhancement of the environment for defence and to harm opponents was regarded as a key component of both feng shui and martial arts. Mountains and water were essential elements to ensure the proper martial positioning and circulation of energy to help achieve success. In Japan, for example, it was believed necessary to appreciate local geomantic configuration to achieve success. Linked to this, iconographic warfare was important in China, notably in the location of defensive walls (the extent to which this was also true for other societies has attracted insufficient attention). Maps from the Chinese Ming dynasty (1368–1644) often have distinctive markers and symbols that denote cultural affiliations and power relations, such as drawing huts for aboriginal peoples and putting walls around cities.

OTTOMAN EMPIRE

Mapping activity from a different governmental and cultural basis was seen in the Ottoman Empire (later Turkish Empire), which by 1534 ruled from its base at Constantinople a territory stretching from Iraq to Hungary and Algiers to Crimea and would control this area until at least the 1680s. Albeit without the institutional and other traditions of the Chinese, the Ottomans used frontier surveys to provide information on key areas in their porous frontiers. The first survey, of the province of Buda, was compiled in 1546 and reflected the military sensitivity of the Habsburg-Ottoman frontier in Hungary. Maps, including siege maps, also served the Ottomans as tools of military reconnaissance and intelligence.

FIFTEENTH-CENTURY DEVELOPMENTS

In Christian Europe, there were significant developments in mapping from the fifteenth century, notably the application of mathematical proportionality to the known world and the impact of printing. The linear perspective, which became important in Western painting from the fifteenth century, mirrored cartography in its attempt to stabilise perception and make it more realistic. In both paintings and maps, there was an emphasis on accurate eyewitness observation, and on observation that was faithfully reproduced. The use of mathematics to order spatial relationships provided a visual record of measured space. In place of idealised and formulaic representations came a desire for topographic specificity, in short an understanding and presentation of difference.

Western advances in trigonometry and, critically, in the dissemination of practice and perception across a broad range were intertwined with the ability to use maps and to understand spatial dimensions without necessarily seeing the physical object. More broadly, humans were stimulated to learn and visualise more, and a self-reinforcing link was established between these processes and book- and map-learning. This situation was linked to the application of knowledge as a process responsive to changing information. There

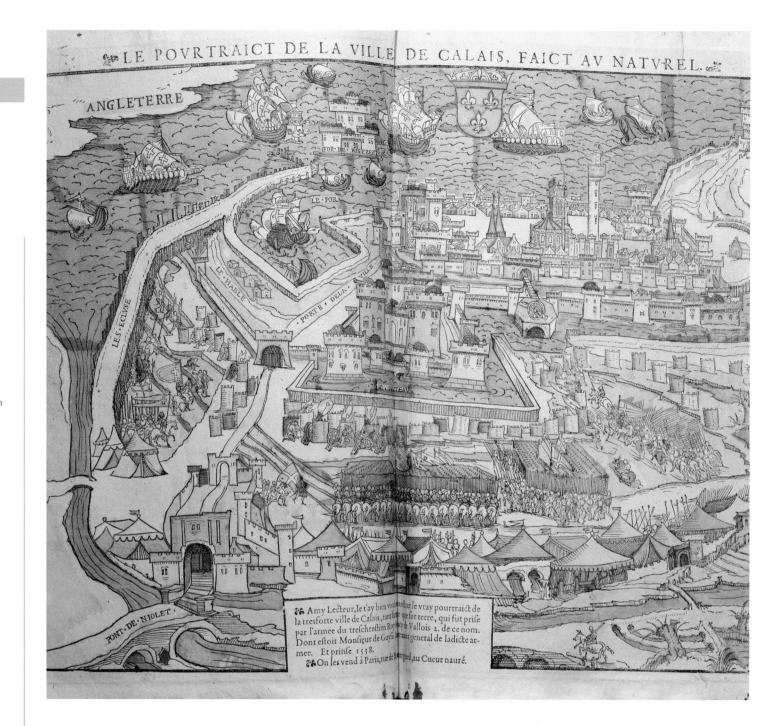

LE POVRTRAICT DE LA VILLE DE CALAIS, FAICT AV NATVREL.

SIEGE OF CALAIS, 1558. FRENCH SCHOOL, SIXTEENTH CENTURY Captured after a long siege in 1347, Calais was England's last position on the French mainland. England went to war with France because of the conflict between the latter and Mary Tudor's husband, Philip II of Spain. In January 1558, French forces bombarded Calais into surrender in a winter campaign characterised by bold French generalship and an unexpected attack. François, 2nd Duke of Guise (1519–63), an experienced commander, gave the French a badly-needed victory at a time of great pressure from Spanish forces. RIGHT

was a willingness to use new knowledge to challenge inherited frameworks, not least the strong prestige of the Classical tradition. Moreover, a map presented knowledge at a distance, and did so in a fixed form and one that was readily understood by more than one observer and over a period of time.

Voyages of exploration became important for Christian Europe from the fifteenth century, starting with the Portuguese venturing down the west coast of Africa, and created suggestions of vast wealth and great power through journeying forth. This process encouraged the official accumulation of cartographic information, notably by the Portuguese at Lisbon, as well, crucially, as attempts to keep it secret. It was illegal to possess charts and globes that had not been approved.

SIXTEENTH-CENTURY EXPANSION

This was mapping for the wars of expansion that took the power of Europe to territories Europeans had never heard of hitherto. Indeed, European maritime hegemony from the sixteenth century rested in part on cartographic developments in Europe that permitted the depiction of the world's surface on a flat base in a

**SIEGE OF SZIGETVÁR, 1566. ITALIAN SCHOOL,
SIXTEENTH CENTURY** In 1566, Süleyman the
Magnificent, on his last campaign,
besieged the fortress of Szigetvár in
southern Transdanubia (western
Hungary), which had been unsuccessfully
besieged a decade earlier. This was a
relatively minor position, certainly
compared with Vienna, which was the
original target. Angered by Count Miklós
Zrinyi's attack on his flank, Süleyman
decided to attack his fortress of
Szigetvár. However, he was held up by
the strong resistance under Zrinyi
(1508–56) who, although heavily
outnumbered, held out from 5 August to
8 September. Running out of food and
ammunition, Zrinyi led a final sortie on
8 September, dying at the head of his
troops. Süleyman (1494–1566) had died
in his camp two days earlier. That this
campaign did not lead to major gains
reflected the Habsburgs' ability to
strengthen their frontier. On the other
hand, the vitality of the Ottoman system
at the close of Süleyman's life contrasts
with the difficulties faced by Philip II of
Spain in the Low Countries in 1566 as the
Dutch Revolt began. LEFT

manner that encouraged the planned deployment and
movement of forces. Thus, cartography was a crucial
aspect of the ability to synthesise, disseminate, use and
reproduce information that was important to
European hegemony. The movement of ships could be
planned and predicted, facilitating not only trade, but
also amphibious operations, for example the Spanish
relief of Malta from Ottoman attack in 1565. Maps
served to record and replicate information about areas
in which Europeans had an interest, and to organise,
indeed centre, this world on themes of European
concern and power.

Maps were particularly important for the
employment of artillery. This was not so much at the
tactical level, because of the problems of recording and
mapping height and, in the beginning, of a limited
range for cannon. Moreover, line-of-sight fire was
coordinated by eye. Instead, maps were important at
the operational level, because they provided
indications of where artillery could be transported.
Maps could be inadequate in their depiction of roads,
but there was an awareness of the need for information
to aid troop transport by the 1530s, and it was
increasingly catered for. In England, Thomas Elyot

praised the utility of maps in *The book named the governor* (1531). More generally, the requirement for information on where cannon could be transported was a product of the operational issues posed by the greater scale of war, and by the need, despite this scale, to retain mobility. Thus, an understanding of routes was crucial to war, and maps provided a key aspect of this.

More specifically, military purposes, notably fortification, accounted for the mapping of certain localities. The increased role of artillery demanded the planning of defensive sites, the better to deploy optimum firepower to cover enemy attack from all sides. Artillery encouraged investment in fortifications, investment that was knowledge-driven. For example, Cesare Borgia (c. 1476–1507), the illegitimate son of Pope Alexander VI and the successful and feared Captain-General of the papal army in 1499–1502, asked Leonardo da Vinci, a gifted physical geographer adept at surveying among his polymathic skills, to examine the papal fortresses. Da Vinci provided accurate maps, for example of the city of Imola in 1502, which were useful tools for planning defences.

Similarly, the first maps drawn to scale by English map-makers date from about 1540 and related to attempts to improve the defences of Guines. This was part of the English defensive system covering Calais, which the English, having captured it in 1347 after a long siege, held until it fell to a surprise French attack in 1558.

The geometric style of fortification encouraged the use of plans and of drawing to scale. Developed from the 1450s, the polygonal bastion was spread across Europe by Italian architects. In a new defensive system known as the *trace italienne*, bastions – generally quadrilateral, angled and placed at regular intervals along all walls – were introduced to provide effective flanking fire. Battista della Valle's work on the subject, *Vallo libro continente appertinente a capitanii, ritenere et fortificare una città con bastioni,* went through eleven editions from 1524 to 1558.

Most positions, however, were not fortified at the scale and expense of the best-defended locations. Moreover, alongside the desire for geometrically perfect shapes, a desire that matched the Chinese commitment to *feng shui,* fortification techniques had to be adapted to terrain and topography. These goals, and the cost of such building projects, made the ability to produce accurate scales crucial in depicting plans. Galileo's first publication, *Le Operazioni del compasso geometrico e militare* (1606), focused on military engineering and emphasised the importance of applying mathematical rules.

Firing lines and ballistics became central issues. In Shakespeare's play *Othello* (1604), Iago complains that he has been passed over for promotion because his rival Cassio has the skill of a mathematician. This related to the need for such skills for the Venetian state: the general, Othello, is presented as sent to Cyprus to defend this Venetian colony against Ottoman attack. In practice, Cyprus had already fallen to just such an invasion, in 1570–71, with the sieges of Nicosia and Famagusta proving the crucial episodes. Famagusta held out for eleven months despite the request for surrender being accompanied by the head of Nicolò Dandolo, the Lieutenant-General of Cyprus, who had been killed at Nicosia.

Maps were also used to plan, assist or explain the movements of military units. Thus, the ultimately unsuccessful English invasion of Scotland in the 1540s was accompanied by mapping, while maps of Franche-Comté, in what is now eastern France, were produced in order to help plan the route for Spanish units moving through the province en route between Italy

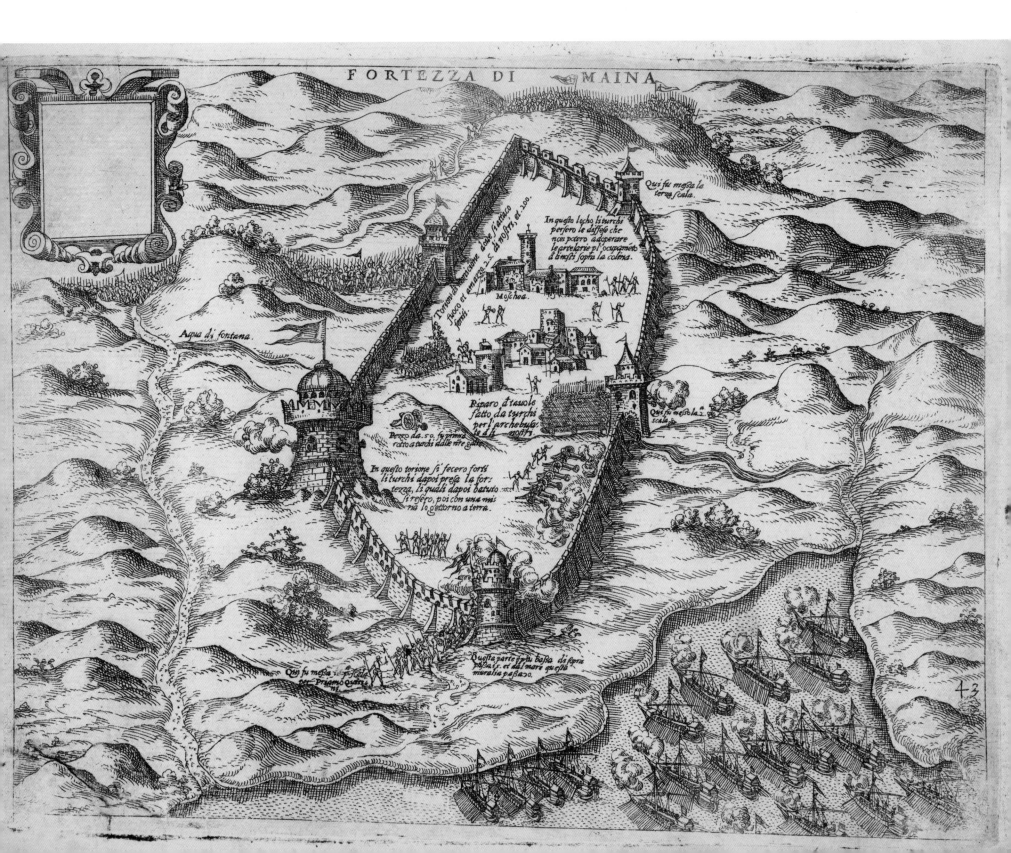

FORTEZZA DI MAINA

and the Low Countries, specifically Genoa, Milan and Brussels, in the late sixteenth century. The maps showed relief features that affected routes and timing along part of the so-called 'Spanish Road'. As the Spanish army was the most impressive and successful in Christian Europe, this usage reflected best practice.

In the late 1540s, as a means of persuading his friends to join him in revolt against his brother, the Protector, Edward, Duke of Somerset, Thomas Seymour tried to use a manuscript map of England to demonstrate the extent and strategic nature of his support. This attempt led to his execution.

In response to the crisis of war with Spain in 1585–1604, another English map, of Lancashire, depicted the seats of both Catholic and Protestant gentry. This map of Lancashire was owned by Elisabeth I's effective leading minister William Cecil, Lord Burghley, and clearly indicated the security problems believed to be posed by the loyalties of the Catholic gentry.

The maps that can still be consulted suggest many more maps were used but did not survive.

PRINTED MAPS

Maps for particular operational purposes were produced by hand. In contrast, the maps that provided background information and conveyed a sense of space were printed. Indeed, maps were first printed in Europe in the 1470s. Although manuscript maps continued to be produced and remained important to the military, printing became central to map production. This change ensured that maps were available in greater quantities, and more recently than hitherto. Printing thereby facilitated the processes of copying and revision that were so important for map-making.

Improvements in production, with woodblocks giving way to engraved copper plates from the mid-sixteenth century, and the change from the screw press to the rolling press, meant that maps were produced more speedily and with greater uniformity. These developments both encouraged and reflected increased carto-literacy, which opened the way to the increased use of maps in the seventeenth century.

FIREARMS AND FORTIFICATION

The changes in warfare in the sixteenth century also reflected the much greater use of a newish technology: that of gunpowder. Although of long-standing use in China, where a formula for manufacturing it had been worked out in the ninth century, and used for cannon in Christendom in the fourteenth, the deployment of large infantry forces primarily equipped with firearms was really a development from the early sixteenth century. As with mapping, the improvement and application of gunpowder brought together a number of elements, including changes in the composition and manufacture of gunpowder, as well as in the metal casting employed for making cannon. The replacement of stone by iron cannon balls, improvements in the transport of cannon and the development of the trunnion (a supporting cylindrical projection on each side of the cannon barrel), which made it easier to change the angle of fire, all increased the effectiveness of artillery. They also posed the formidable logistical challenge of transporting not only the cannon but also the cannon balls and powder.

At the same time, firearms were integrated as part of defensive/offensive formations that included troops armed with cold-steel weapons, notably pikes. This integration placed a premium on command ability, and that was part of the process in which spatial skills became more significant.

An emphasis on the use of visual images when planning fortifications and, linked to this, their role in

JULIUS CAESAR CROSSING THE RIVER RUBICON. BY EGNAZIO DANTI, 1536–86 The sense of the Classical world as separate and different was not to the fore in the Renaissance. Instead, modern and ancient warfare were regarded as similar and were presented accordingly. The sumptuous display of maps of Papal possessions in the Galleria delle Carte Geografiche in the Vatican painted by Egnazio Danti included this depiction of Julius Caesar crossing the River Rubicon in 49 BCE in his successful grasp for power. Caesar's army was presented as if it was a modern force. Caesar was widely seen as a model for current generals. Andrea Mantegna's series of paintings *The triumphs of Caesar* (1482–92), for example, were displayed for Francesco II, Marquess of Mantua, as a comparison to his family's military history. Caesar's *Comentarii* were published in an Aldine edition in 1513 and, again in Latin, in Paris in 1543. More generally, the Classical world was regarded as a crucial fund of knowledge. In his *Art of War* (1521), the Florentine thinker Nicolò Machiavelli tried to update Vegetius' *On Military Matters* (dating to the fourth or fifth century CE) by focusing on the pike and treating the handgun as similar to missile weaponry. **LEFT**

William Cecil, Lord
Burghley, Elizabeth I of England's
secretary and most prominent minister,
was a key supporter of cartography. A
prominent figure in a developing map
culture, he commissioned, used and
displayed maps. His country house at
Theobalds contained in 1591 in the gallery
a large wall map 'of the Kingdom, with all
its cities, towns and villages, mountains
and rivers; as also the armorial bearings
and domains of every esquire, lord,
knight and noble who possesses lands
and retainers to whatever extent'.
Burghley was a major patron of the
leading English cartographer Christopher
Saxton and collected his county maps.
Lancashire, the English county with most
Catholics, was a centre of governmental
concern about insurrection. In seeking to
assess the local political situation,
Burghley found maps of great value.
There was to be no rising in Lancashire to
follow the unsuccessful one by the
Northern Earls (Earls of Northumberland
and Westmorland) in 1569. RIGHT

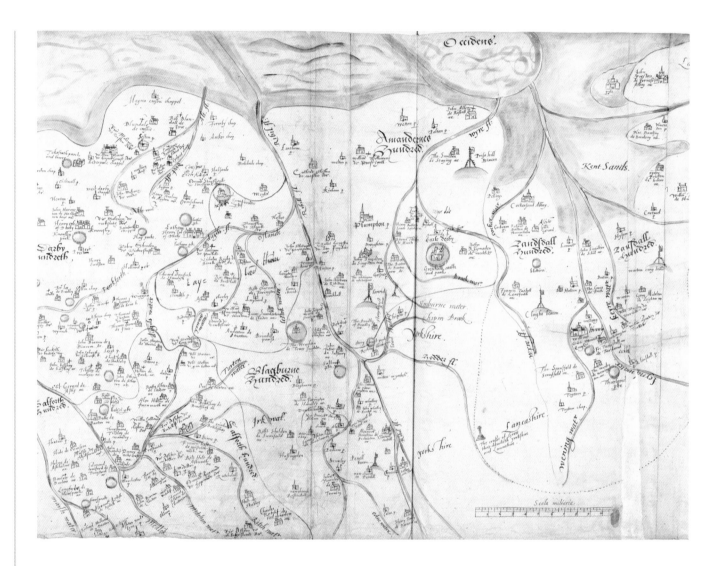

the intertwined practices of geometry and ballistics
ensured that diagrams became more common.
Diagrams also became more significant as
publications were increasingly illustrated. These
publications recorded and encouraged the process
by which a self-conscious method of rational
analysis was applied in the European discussion
of warfare and weaponry. The use of diagrams
encouraged that of maps.

The best ways to organise defences and use cannon
were clear instances of such rational analysis. Girolamo
Cattaneo (d. *c.* 1584), who settled in the Venetian city
of Brescia in 1550, teaching military matters, was a
prolific writer on this subject. His works, including
Opera nuova di fortificare, offendere, et difendere (1564)
and *Nuovo ragionamento del fabricare le fortezze; si per
prattica, come per theorica* (1571), were frequently
reprinted and he was consulted by Vespasiano Gonzaga
about the transformation of the fortress of Sabbioneta
into a town.

Der honige Nauar sfolck

PARIS

Von nbargh

Mar Von Rmti

P. Vo chimy

P. Von Parma

Der h' von barbanson

Graff Vo Barlemont

G. Von Bossu

SIEGE OF PARIS. BY FRANZ HOGENBERG, 1590 In 1590, having gained the initiative by a victory over the forces of the Catholic League under the Duke of Mayenne at the battle of Ivry, Henry IV, the Protestant claimant to the throne of France, besieged Paris, although he lacked a strong enough army to dominate the situation.

The privations of the siege were harsh, but Paris was relieved by the Army of Flanders under the talented Alessandro Farnese, Duke of Parma, who with 14,000 troops advanced from the Netherlands to join the forces of the Catholic League. Henry tried to engage Parma in battle, but he was outmanoeuvred. Parma,

who was strong in infantry, used entrenchments to block Henry's advantage in cavalry. The siege was abandoned. Henry did not enter Paris until 1594, and that success owed much to his conversion to Catholicism. ABOVE

THE DUKE OF PARMA RELIEVING ROUEN. BY FRANZ HOGENBERG, 1592 Henry IV, the Protestant claimant to the throne of France, besieged Rouen in November 1591, joining an English force under Elizabeth I's favourite, Robert, 2nd Earl of Essex, concerned that such positions on or near the English Channel could serve as invasion bases for another Spanish Armada. Attacks, however, were repulsed and the Spanish Army of Flanders under its commander, Alessandro Farnese, Duke of Parma, relieved the city. In this operation, the Spaniards built a bridge of boats across the River Seine, demonstrating the cohesion of the organisational structure of the Army of Flanders and its high level of training. However, Henry was helped by his enemies' political divisions and by his willingness to become a Catholic again in 1593. Resistance crumbled in 1594–96.

RIGHT

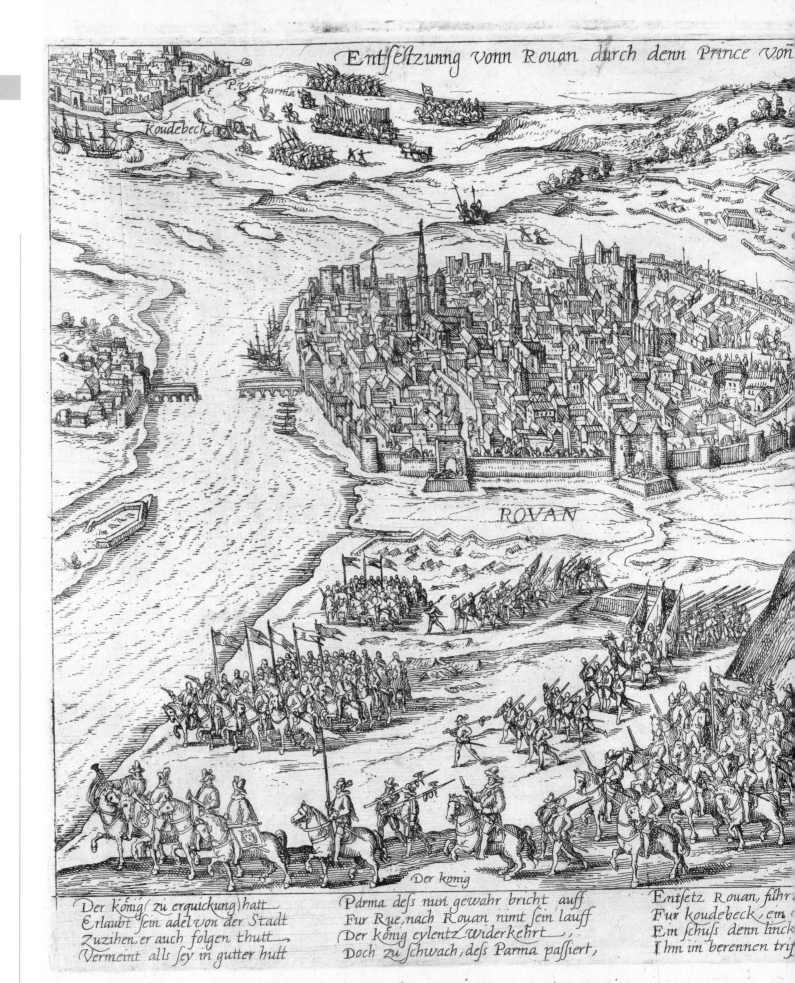

Entſetzung vonn Rouan durch denn Prince von

Koudebeck

P. v. parma

ROUAN

Der konig

Der konig (zu erquickung) hatt
Erlaubt ſein adel von der Stadt
Zuzihen, er auch folgen thutt,
Vermeint alls ſey in gutter hutt

Parma deſs nun gewahr bricht auff
Fur Rue, nach Rouan nimt ſein lauff
Der konig eylentz Widerkehrt,
Doch zu ſchwach, deſs Parma paſsiert,

Entſetz Rouan, führ
Fur koudebeck, ein
Ein ſchuſs denn linck
Ihm im berennen tri

L'arrivee de l'armee du Duc de Parme

The concern with fortifications could be seen in works such as *I quattro primi libri di architettura* (Venice, 1554) by the architect and mathematician Pietro Cataneo; *Discorsi di fortificationi* by Carlo Teti (Rome, 1569), who taught the art of war to the future Maximilian of Bavaria; and *Delle fortificationi* (Venice, 1570), by Galasso Alghisi, architect to the Duke of Ferrara. Antonio Lupicini, a military architect and engineer, published *Architettura militare* in Florence in 1582.

Italy was not alone. Thomas Styward's *The pathway to martial discipline* (1581) was copiously illustrated with battle plans. When Thomas Morgan returned to England, having served in the Low Countries against the Spaniards in 1572, he produced an illustrated account that showed in considerable detail the tactics and difficulties of operating in such a water-girt landscape. Paul Ive's *The practise of fortification* (1589) devoted particular attention to Dutch-style fortifications. With its preference for illustrations, the world of print proved conducive to the development of maps as a form of both educational diagram and attractive illustration. The frequency of wars in Europe in the period encouraged this process and it was further enhanced by the keenness with which practitioners turned to print. Thus, Buonaiuto Lorini, a Florentine who designed fortifications for Venice, France and Spain, published *Delle fortificationi* in Venice in 1596. The entrepreneurial character of publishing was important for the appearance both of such works and for an increasing number of printed diagrams, plans and maps.

RELIGIOUS DIMENSIONS

Religion is an underrated dimension of the topic of maps and war, because the mapping did not involve

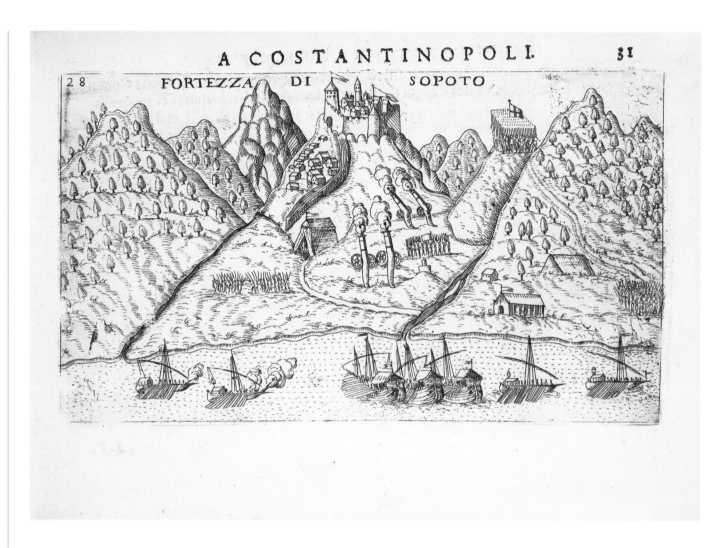

what is seen as a realistic cartography. However, the idea of humans as occupying a world defined as well as created by spiritual forces was matched by a sense that the essential parameters had a real physical existence. If this was true of heaven and hell, that reflected the belief that the agents of good and evil had bases and a real presence. Their continual rivalry was the key war, and one, moreover, that shaped the struggles between humans. As a result, there was mapping of the existential struggle between good and evil. In large part, this struggle was depicted in the traditional

artistic presentations of the Last Judgment, with the good being raised to heaven and the evil dragged down to hell. For Protestants, with their emphasis on literary discussion, rather than visual depiction, there was also, building on earlier sermons, an admonitory literature that looked at the choices made as Christians sought to achieve salvation. The resulting descriptions/illustrations did not accord with modern ideas of maps, but they were of considerable importance in suggesting one way in which strife and space were conceptualised and conveyed. The pictorial element was much to the fore in this field.

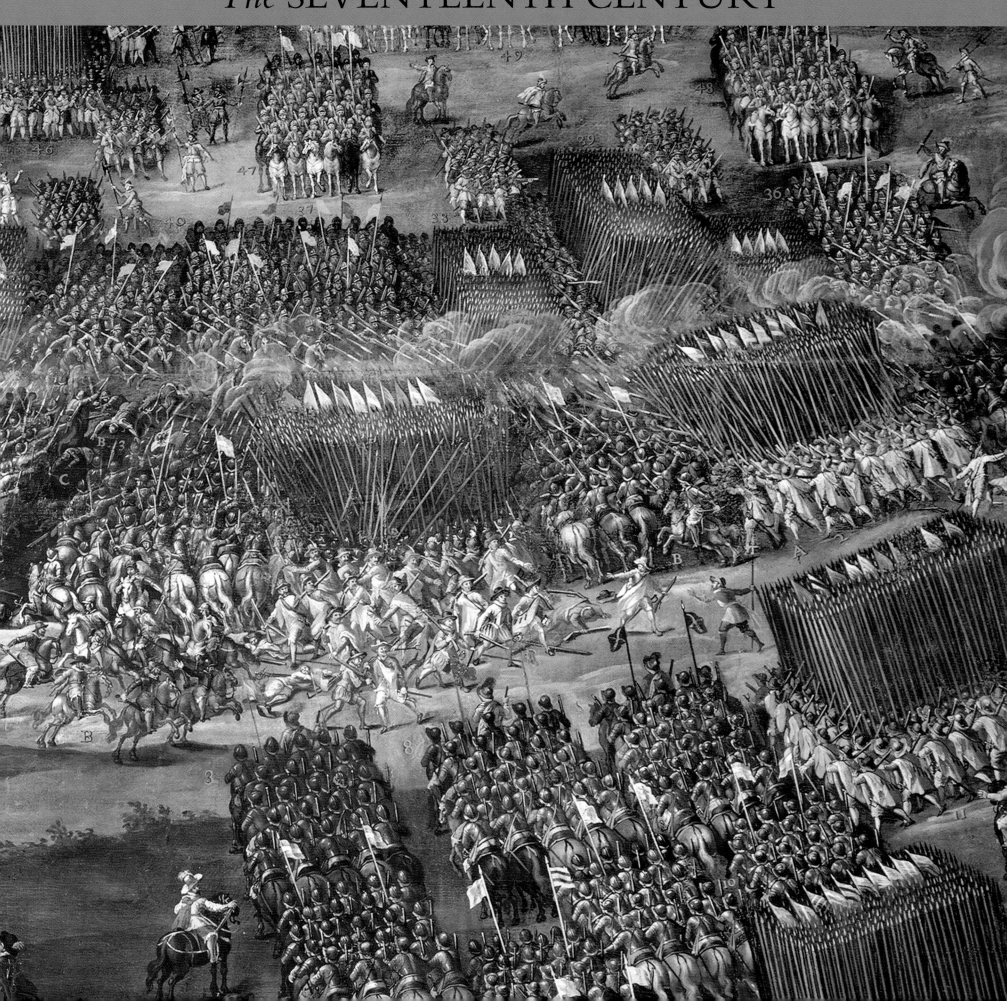

MAPPING *for* SIEGES

ATTACK ON A FORTIFIED TOWN, FROM *TRAITÉ DES SIEGES*. BY SÉBASTIEN LE PRESTRE DE VAUBAN, **1633–1707** A master of fortification and siege craft, Sébastien Le Prestre de Vauban (1633–1707) had a high reputation that was enhanced and sustained by the publication of his works well into the eighteenth century even outside France. In the *Traité des Sieges*, Vauban argued that the greater number of fortresses placed an increased premium on siege craft. In the defence, Vauban's skilful use of the bastion and of enfilading fire represented a continuation of already familiar techniques, especially layering in depth, but it was the crucial ability of France to fund and implement such a massive programme that was novel. In the attack, Vauban showed in the successful siege of Dutch-held Maastricht in 1673 how trenches could march forward by successive parallels and zigzag approaches, designed to minimise exposure to artillery and sorties, and to advance the positions from which besieged fortifications could be bombarded and attacked. Other successful sieges included Valenciennes (1677), Ypres (1678), Mons (1691), Namur (1692) and Ath (1697). The success of sieges ensured that battles became more significant as a way to block them. RIGHT

THE BATTLE OF WHITE MOUNTAIN 1620. BY PIETER SNAYERS (1592–1667) The battle took place near Prague on 7–8 November 1620. PREVIOUS PAGES

MAPS PLAYED a more significant role in war, and the discussion of war, in the seventeenth century in Christian Europe than they had done hitherto. This increase was as part of the developing role of printing in knowledge about war. Printing, moreover, was important in strengthening the consciousness of a specific European military tradition, not least as printed manuals on gunnery, tactics, drill, siegecraft and fortification spread techniques far more rapidly and, generally reliably, than word of mouth or manuscript had done. Manuals also permitted the sharing of knowledge within a context of standardisation that, at least in the long term, both helped to increase military effectiveness and was important for cohesion and the use of military resources.

Maps were an important part of this fixing and sharing of information. Alongside printing and literacy, they fostered discussion of military organisation and methods, and encouraged a sense of system, affecting and reflecting cultural assumptions. Maps were frequently referred to in the arts. For example, in the *Portrait of a Man Reading a Coranto* (c. 1675) by Gerard ter Borch (1617–81), the interior scene depicted a book or atlas open on the table with a double-page map showing. Similarly, in the *Interior of a Study* (c. 1710–12) by Jan van der Heyden (1637–1712), there was a globe and an atlas or book containing a double-page map.

Nevertheless, it is necessary to avoid a triumphalist account that automatically suggests a major and successful transformation. For example, there was no automatic transition from printing to the large-scale use of military mapping. Furthermore, the process of change was far more apparent in Europe than in China where printing was invented. At the same time, there is a lack of knowledge about how Chinese

ATTAQUES D'UNE PLACE entourée de Fausses Brayes

A,B, Bastions du front de l'attaque.

C, Demy Lune du meme front.

D, fausses Brayes.

E,F, Demy Lune Collateralle.

AG,BG, Prolongemens des capitales des bastions A,B,

I, Piquets sur la lignement des capitales garnis de paille ou de Meche allumée pour servir a la conduitte des attaques.

K, Batteries a ricochets des deux faces et du chemin couvert de la demy lune C,

L, Batteries a ricochets des Bastions A,B,et de leurs fausses Brayes.

M, Batteries a ricochets des faces et chemins couverts des demy lunes collateralles E,F, qui voient sur les attaques

N, Batteries a Bombes.

O, Places sur la deuxieme Ligne ou l'on pourroit Metre les Batteries a ricochets et a bombes s'il etoit necessaire de les changer.

P, Demy places D'Armes

Q, Cavaliers de Tranchée qui Enfilent le chemin couvert.

R, Passages de fassines pour Mener les Canons et mortiers a leurs Batteries.

Echelle de 200. Toises

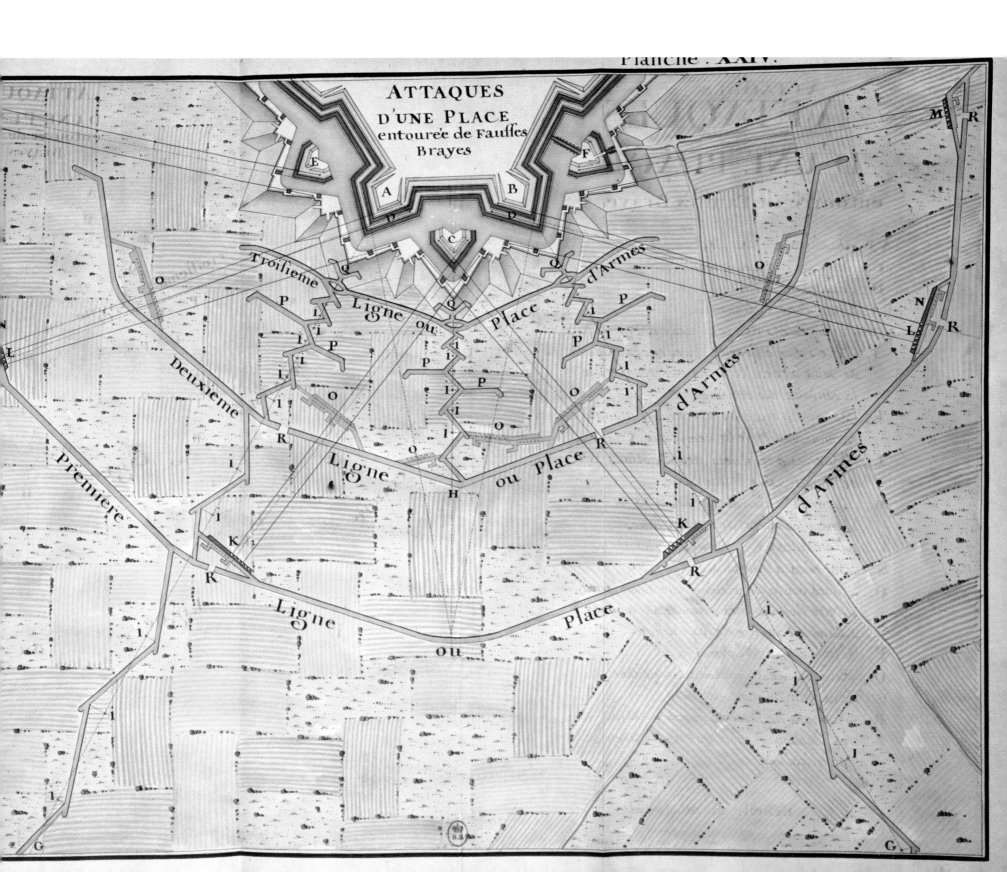

Planche XXIV.

ATTAQUES
D'UNE PLACE
entourée de Fausses
Brayes

OCEANVS GER MANICVS

The Princes Battel:

The Arch-Dukes Batte

NEVPORT.

BATTLE OF NIEUPORT, 1600. ENGLISH SCHOOL, SEVENTEENTH CENTURY At once a picture and a plan. Advancing on Ostend, the major privateering base, the 12,000-strong Dutch army, under their commander in chief, Prince Maurice of Nassau, was caught by the 11,500-strong Spanish army under the Archduke Albert of Austria at Nieuport. Albert was the husband of the Infanta Isabella and they were joint rulers of the Spanish Netherlands. By making extravagant promises, Albert persuaded his mutinous veterans to return to duty. The mutinous condition of the Spanish field army had been one of the main reasons for Nassau attempting the risky enterprise. It was not anticipated that the Spanish forces could be brought back to the colours so quickly, and doing so says a lot for Albert's leadership. After holding off Spanish attacks, the Dutch launched a cavalry assault that broke the opposing infantry. However, Dutch losses lessened the value of victory because it was difficult to replace veterans. Discouraged by the approach of Spanish reinforcements, Nassau withdrew. Nevertheless, the battle demonstrated that the Spanish army could be defeated in the open field. The dynamics of the battle could not be shown on the map. ABOVE

SIEGE OF BREDA, 1625. BY JACQUES CALLOT, 1627–8
Captured by the Dutch under Maurice of Nassau in 1590 and by Spain in 1625, after the garrison was starved into submission, Breda fell anew to the Dutch in 1637. These respective fates reflected wider developments. In particular, the Dutch benefited from France going to war with Spain in 1635. However, each position still had to be besieged, and this represented a formidable effort. The sieges were followed with great attention by foreign audiences. Herman Hugo, the chaplain to Ambrogio Spinola, the commander of the Spanish Army of Flanders, was present at the latter's successful siege of Breda in 1625, and published an account of it in 1626. This illustrated account appeared in Latin, with a second edition in 1629, and was translated into English (twice), French, Italian and Spanish. It influenced Jacques Callot who was commissioned by the Infanta Isabella, Regent of the Spanish Netherlands, to commemorate the siege. Callot visited Breda at least three times and depicted himself at work in his etching which combined episodes from the siege in one image. The map took several approaches, with a picture at the bottom, a plan in the middle and an elevation view at the top. Breda was allocated to the Lands of the Generality and remained Dutch thereafter. **LEFT**

commanders used maps, while, similarly, it is unclear whether the greater availability of printed material in Christian Europe in the seventeenth century indicates that it had a major impact on operations. There was a lack of standardisation in maps. In particular, the absence of a standard system of coordinates made difficult the communication to another party of the spatial information conveyed by a map.

This issue does not only relate to the strengths and deficiencies of maps and mapping. In particular, the selection of commanders put more of an emphasis on birth than on education. Moreover, military education was largely on the job, and there were few military academies. Therefore, there was no formal education in map use. In addition, the nature of command, with a marked preference for verbal orders, as well as the limited character of the sources, make it very difficult to establish the precise extent of map usage. The situation was to be

very different by the late nineteenth century when maps played a major role in the training of senior officers in the General Staff System.

Despite its limitations, map usage in Christian Europe developed during the seventeenth century and the preface to the *Atlas historique* (1705) felt able to claim, 'The map is a help provided to the imagination through the eyes'. In what was to become a key pattern, maps appeared both in response to events, informing the public, and in a prospective fashion, preparing states for war. A classic instance of the former was the map published in The Hague in 1635 depicting Dutch conquests from Portugal's major colony, Brazil. It provided both details of particular sites, such as the major city of Recife (captured in 1630), and a large-scale map of Brazil on which the general campaign could be followed. As a result, key elements, particularly sites and the region of contention, were combined on that map. No comparable map was produced for the Dutch loss of

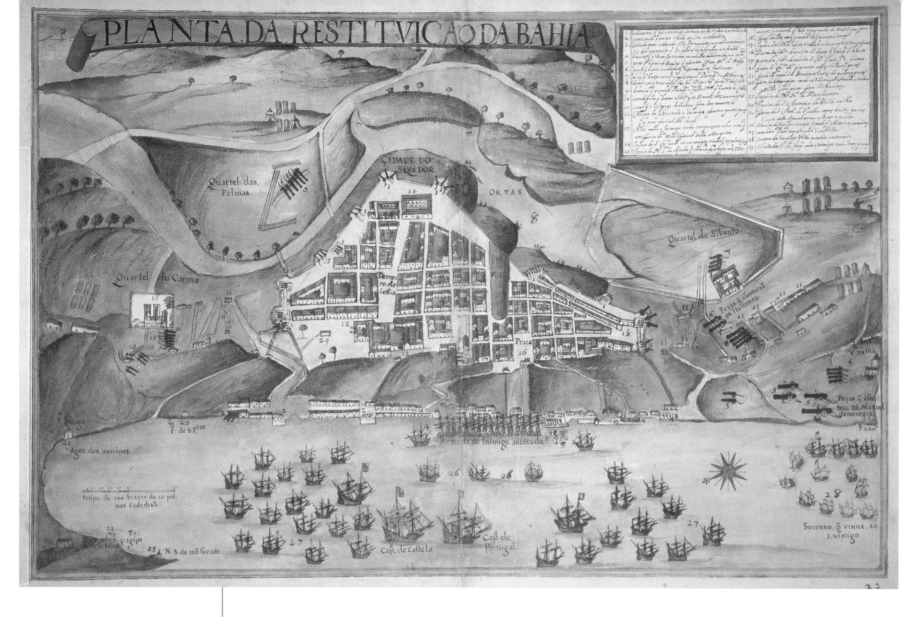

THE PORTUGUESE LIBERATION OF THE CITY OF

SALVADOR, BRAZIL, 1631. BY JOÃO TEIXEIRA

ALBANAZ Despite initial success and
major efforts, the Dutch failed to drive
the Portuguese from Brazil, a major
source of sugar for Europe and a key
slave economy. In 1630, they established
New Holland, which covered much of
north-eastern Brazil, but in 1645 their
position was challenged by a large-scale
rebellion, and the subsequent attempt to
stabilise the situation by capturing the
capital Salvador (Bahia) failed. From 1652,
the Dutch focused on war with the
English and, in 1654, the Portuguese
re-established their control in Brazil.
They also drove the Dutch from their
gains in Portugal's African colonies. ABOVE

their positions in Brazil from 1645. This process was
completed in 1654 when Recife was recaptured by the
Portuguese, a key event in their mid-century revival.

Frequently, however, these elements of scale and site
were separate. Thus, in 1689, Thomas Phillips
published plans of the city of Londonderry in northern
Ireland which was then being blockaded on behalf of
James II (VII of Scotland). In the event, William III's
forces broke the siege, with the English fleet breaking
the boom blocking the harbour. Two years earlier,
Giovanni Chiarello's history of the Austrian campaigns
against the Turks in Austria and Hungary in 1683–86,
campaigns that began with the defeat of the Turkish
force besieging Vienna, included a number of maps.
That of the Austrian siege of Nayhaysel in 1685
depicted bastions, trenches, cannon and troops. It also
offered a good guide to the shape of the siege,
although, as with other maps of sieges, not one of its

progress. The extensive key, however, provided
substantial detail.

MAPS AND PICTURES

That plans were produced was an instructive variant
on the earlier focus on pictures, although the latter
remained prominent in recording events and, possibly,
in sketching plans. Furthermore, there was the
important fusion type of map: an illustration of the
conflict that included a map of the battle. In this
format, the illustration provided the dynamic quality
and sense of vigour, and an aspect of the visual appeal;
it could also throw light on the topography of the
battle. The map, however, was clearly part of the
authority of the scene. Jacques Callot's map *The siege of
Breda*, produced in 1627–28, depicted the successful
1624–25 Spanish siege of the major Dutch fortress, a
key operation in the war between the two powers

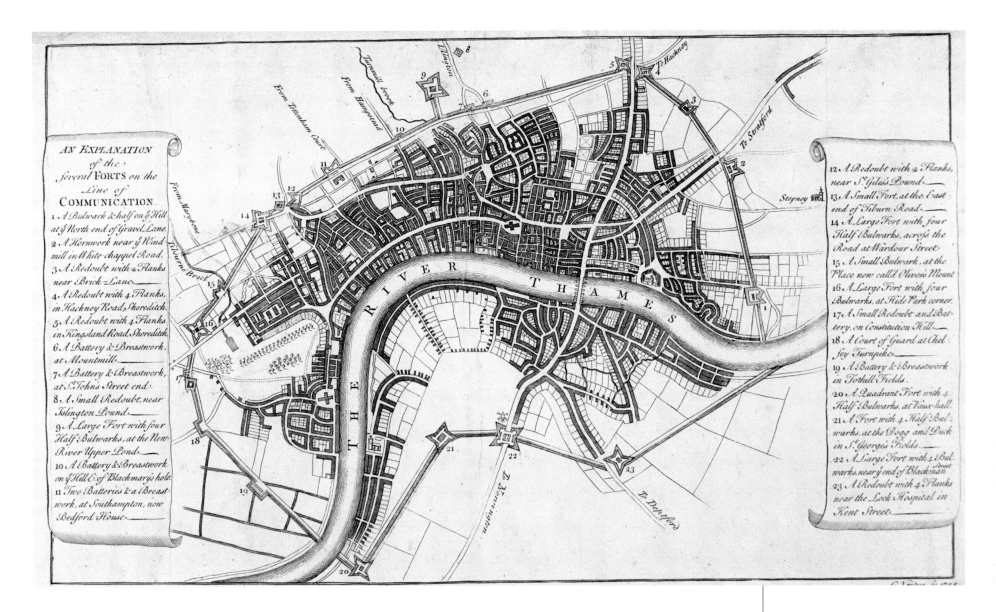

LONDON, AS FORTIFIED IN THE ENGLISH CIVIL WAR. BY GEORGE VERTUE, 1642 The advance of hostile Royalist forces in November 1642 and their sacking of Brentford in Middlesex led to the rapid construction of an eleven-mile-long defence system for London by their Parliamentarian opponents. This was an earthen bank and ditch with a series of 28 forts and two outworks. Sites of forts included what are now the Old Kent Road, the Elephant and Castle, the Imperial War Museum, Vauxhall, Millbank, Hyde Park Corner, Great Russell Street and Great Ormond Street. The area covered was much greater than that of London's medieval walls. Much of the work was completed voluntarily, the Venetian ambassador claiming that 20,000 citizens worked without pay; but the City of London authorities did have to pay for some of the work, and special taxes were raised accordingly. London was the key political, financial, manufacturing and naval centre for the Parliamentarian cause. In the event, a Royalist attack never came, which was just as well as the fall of Bristol in 1643 showed that Parliamentary-held cities could fall to attack. The rapid subsequent removal of the fortifications helped ensure that walls defined London's later street plan far less than in cities such as Vienna. **ABOVE**

from 1621 to 1648. This event had already attracted great attention with Hermannus Hugo's 1626 Latin account of the siege, which had been translated into English, French, Italian and Spanish. The map combined a number of approaches, with a picture at the bottom, a plan in the middle, and an elevation view at the top. The image provided was of public celebration, which was important to the display, perception and thus production of maps. In that respect, maps very much overlapped with pictures, each providing an element of authority for the latter and an aspect of interest for the combined work. Breda was recaptured by the Dutch in 1637 at a time when Spain was under greater pressure as France had entered the war in 1635 as an ally of the Dutch.

The continuum of maps and pictures was such that it is difficult to distinguish between the two categories or to determine how valid modern definitions are. For example, a series of coloured maps devoted to the Venetian conquest of the fortress of Clissa (Klis) near Split (in modern Croatia) in 1648 from the Ottomans, assembled in an 'atlas' dedicated to the Molina family, comprised both pictorial and cartographic elements, the former including troops moving and firing.

MAPS FOR THE FRENCH STATE

Maps for the state were increasingly important during the seventeenth century. Under Henry IV of France (r. 1589–1610), who saw himself as a reformer and was certainly a source of activity, the *ingenieurs du roi* produced detailed manuscript maps of the major frontier provinces. These maps were designed to aid campaigning against Spain and Savoy-Piedmont (both of which Henry fought) and were soon printed.

One of the engineers, Jean de Beins, who was also a fortification expert in the French frontier province of Dauphiné, drew maps based on his surveys of the

different valleys in the Alps which he then linked into a master map. This was an established means of producing a map. These surveys revealed the value of the individual valleys for invasion routes and their vulnerability to attack. The mapping of the Valtelline valley, a key route between Italy and Austria, and one used to move Spanish and Austrian troops, fulfilled a similar function. Such mapping was both of local use and permitted assessment, or at least consideration, at a distance.

The French use of mapping was important because, in the seventeenth century, France became the kingdom that set the model for many other rulers, both in governance and in image. This was notably so under Henry's grandson, Louis XIV (r. 1643–1715), the 'Sun King'. In his early years, before he took personal control of the government in 1661, military mapping was pretty basic. A few good sketches and ground plans of fortifications survive. These were mostly drawn on site, by military engineers. These sketches and ground plans were one-offs because of a particular siege during the war with Spain from 1635 to 1659, and were not part of a systematic programme. As a result, there was no comprehensive basis for consulting maps. Indeed, despite earlier anticipations, military mapping in France largely began in the 1660s.

Other governments feared and/or allied against Louis XIV, but he could not be ignored as a model of kingship. Louis was concerned to map France and to produce maps that would be valuable for war, of which he was very fond. The two goals were not separate, not least as maps of a country were highly important if its defences were to be prepared and its defence planned. They were increasingly believed to be important and French provincial officials were ordered in 1663 to send all available maps and geographical information to Paris. This knowledge was then used

RELIEF PLAN OF CANDIA (HERAKLION) The site of one of the epic sieges of military history. In 1645, the Ottomans invaded the Venetian island colony of Crete, which lay athwart the key grain supply route from the port of Alexandria in Egypt to the Ottoman capital of Constantinople (Istanbul). Sultan Ibrahim sent 348 ships and 51,000 troops, and much of the island was overrun in 1645–46. However, the siege of the capital lasted from 1647 until 1669. The Venetians dispatched reinforcements, including 33,000 German mercenaries, while political divisions undermined the Ottoman war effort, which also faced serious logistical problems. Successful Venetian attacks on Ottoman supply lines hindered the latter in the late 1640s and 1650s, but Fazil Ahmed, the Grand Vizier, personally led a renewed effort to take Candia in the late 1660s, albeit with heavy casualties. The Venetians were greatly outnumbered at sea and the Ottoman fleet cut off the flow of reinforcements from Christian Europe. In 1669, a quarrel between the Venetians and their European allies in the garrison led the latter to retire and forced the Venetians to surrender. This relief plan survives on a Venetian church. It indicates the variety of settings for maps. LEFT

by Nicolas Sanson, the royal geographer, to devise a series of maps of France.

A permanent collection of maps for military purposes, the Dépôt de la Guerre, was founded in 1688. The records of fortifications there included the *Recueil des plans des places du roi*, also known, after the influential minister of war, as the *Louvois atlas*. Compiled in 1683–88, this was an atlas of fortresses. They were the crucial protection of frontiers and the presentation of sovereign control, the latter a key political tool for domestic as well as international politics.

FORTRESSES AND SIEGE WARFARE

As far as fortresses were concerned, maps offered both a pragmatic tool and an emblem of possession. In the latter case, maps acted to record the power and triumphs of the ruling houses of Europe and were part of the resulting panoply of display, alongside paintings, coins and medals. All the visual magnificence was for

display and was on display. Maps were part of a broader pattern of the illustration and discussion of fortifications. However, the role of the latter has been generally underrated due to a preference for discussing battles and, notably, the concept of decisive battles. In practice, sieges were central in many campaigns and, indeed, there was the concept of decisive sieges. So also outside Europe, as in Mughal campaigns in the Deccan, notably in the 1630s and 1680s, or the repeated significance of control over the city of Kandahar to campaigning between the Safavids and the Mughals.

In Christian Europe, there were many publications devoted to the topic of fortification. Antoine de Ville, who became military engineer to Louis XIII of France in 1627, was responsible for *Les fortifications … contenant la maniere de fortifier toute sorte de places tant regulierement, qu'irregulierement* (1628) and *De la charge des gouverneurs des places … un abrégé de la fortification* (1639). His experience and publications were built on by Sébastien

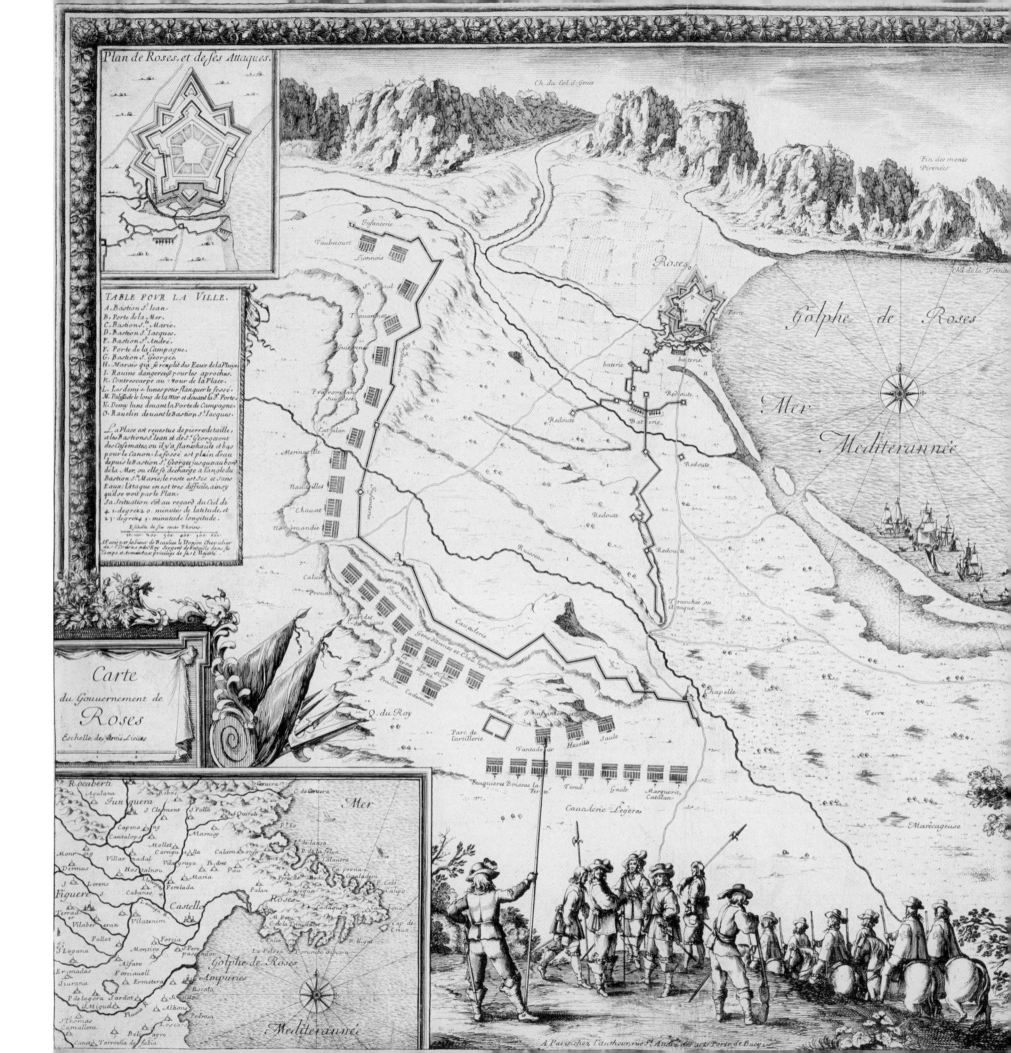

Le Prestre de Vauban, the key builder of fortresses for Louis XIV.

A master of positional warfare, Vauban was the leading European expert in siegecraft. His sieges were recorded in many maps. Vauban showed in the successful siege of Dutch-held Maastricht in 1673 how trenches could more safely be advanced close to fortifications under artillery cover by parallel and zigzag approaches. The garrison of this major fortress capitulated to the French after a siege of less than a month, as did Valenciennes in 1677 and Ypres in 1678. Some positions fell very rapidly – Limbourg in 1676 fell a week after the trenches were opened.

Vauban's reputation led to the publication of his works, and thus plans of the relevant fortresses, well into the eighteenth century. This was true even outside France. The fifth edition of his *New method of fortification* was published in London in 1748, while an edition of his collected works was published in Amsterdam and Leipzig in 1771. The importance of fortifications and sieges was such that others also published on the subject, while translations spread knowledge. Thus, Tomaso Moretti's *Trattato dell'artigliera* appeared in English editions in 1673 and 1683, with the translator, Jonas Moore, also publishing his own *Modern fortification* (1689). As standard practices were adapted to particular circumstances, notably the topography, the specificity of individual fortresses and sieges ensured that detail was very important to the mapping.

The extent to which mapping was involved in the more general question of where best to locate fortifications is unclear in the sense that, as so often with the use of maps, direct evidence is lacking. Nevertheless, it is apparent that this location would have required a spatial space that had to be substantiated through maps. At this point, the general

SIEGE OF ROSAS, 1645 This port city, on the Catalan coast of Spain north of Barcelona, was besieged by French forces in 1645. The French had to guard against a sea-borne relief attempt, but were able to block the Spanish naval operations. With the Comte d'Harcourt and a large part of the French force standing off to the west of Rosas to block any relief by land, Plessis-Praslin's force dug in around Rosas in early April, fought back a sortie by the besiegers a few days later and settled down to the siege. A week later, torrential rain led to damage, demoralisation and a Council of War, where it was nearly decided to abandon the siege. When the rains finally stopped, the besiegers pushed forward the trenches again and settled down to mine the fortifications. As often in the 1630s and 1640s, the siege artillery was relatively modest – ten cannon – and played only a limited role. Mining was far more important, and the fortifications were compromised by a series of successful explosions destroying a couple of bastions by the end of April. The governor agreed to surrender by the last day of May unless he was relieved by then, an unlikely event that did not occur. However, the French failed to sustain their gains in Catalonia. LEFT

maps of countries would have played a role, thus
indicating that they were significant for military
purposes, even when they had no ostensible military
function, again a point of more general relevance.

Fortresses were intended to form defensive systems.
This was seen not only in France but also in Spanish-
ruled Lombardy where the forts of Novara,
Alessandria, Tortona and Valenza were regarded as
links in a chain aimed at the defence of the state from
the west. As a result, the cession of Alessandria to
Savoy-Piedmont in 1713 was a key step that
compromised this system.

Another aspect of the same use of general maps was
provided by civil wars. At some time or other, these
occurred across most of Europe in the sixteenth and
seventeenth centuries, and sometimes for long periods.
Thus, in France, there were civil wars in the 1560s to
1590s, the 1620s and 1648–53, as well as a series of
large-scale rebellions, including in 1636, 1639, 1672
and 1675. In such conflicts, it was useful to have maps
in order to understand developments and to plan or
report accordingly.

SWEDISH AND RUSSIAN EFFORTS

It was not only France that saw governmental efforts
to acquire information and a related bureaucratisation
of spatial knowledge. In the case of Sweden, one of the
most dynamic military powers of the period, there was
a marked change between the sixteenth century and
the first half of the seventeenth century, with the start
of military map-making and naval cartography, both
by collecting maps and by the training of professional
map-makers. The Swedish war archives contain a
collection of maps from the Thirty Years' War (1618–
48, which Sweden, under King Gustavus Adolphus,
entered in 1630) and later, reflecting the army's need
of maps for planning and information during

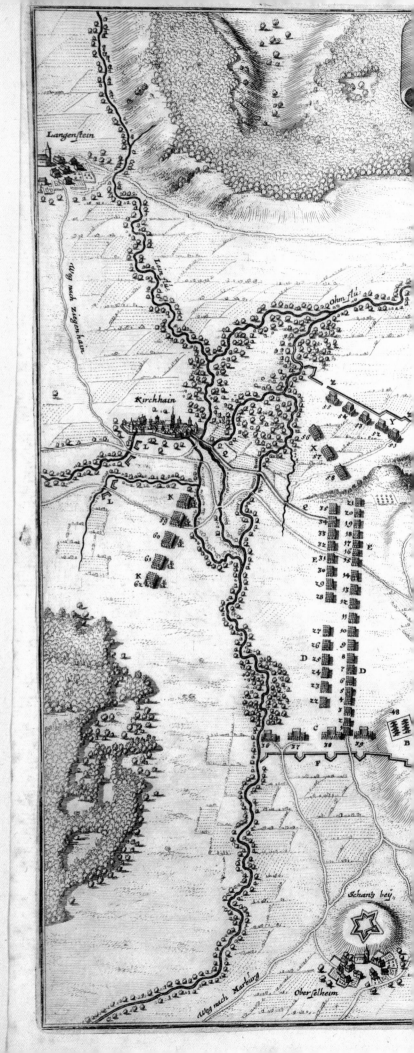

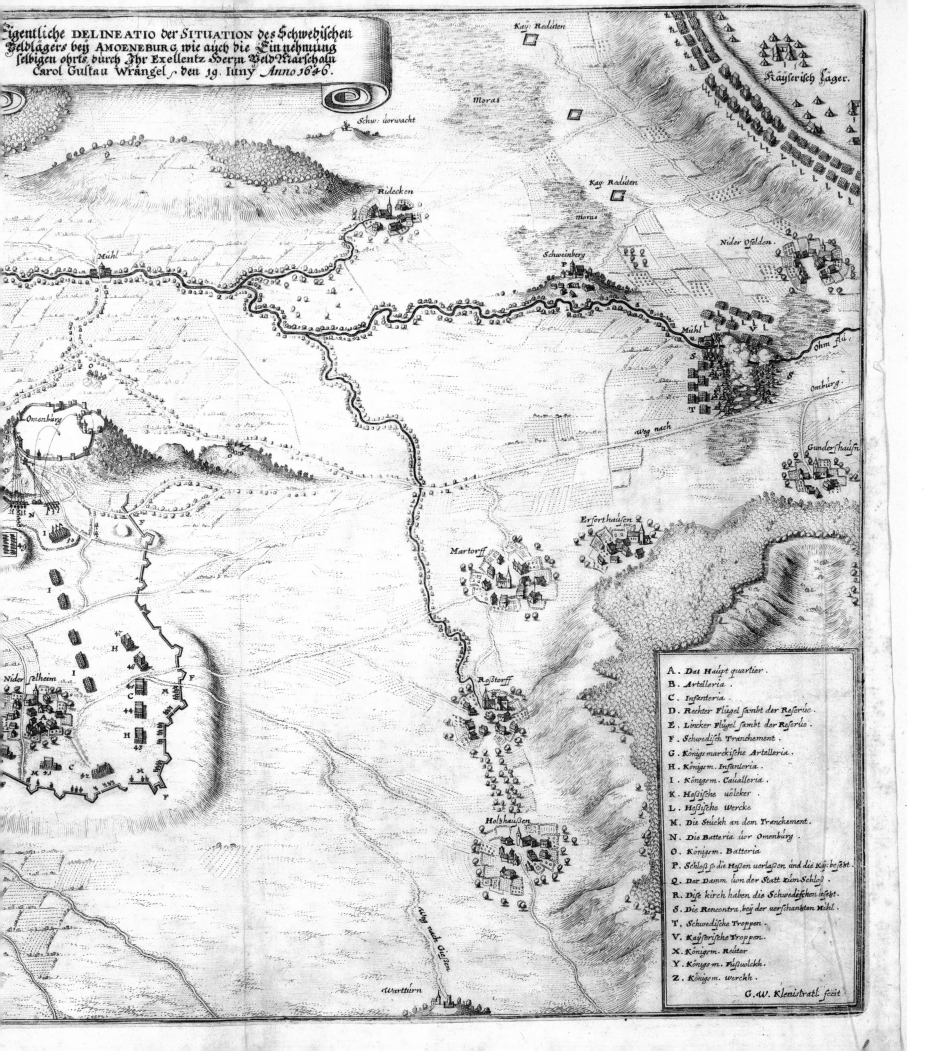

Eigentliche DELINEATIO der SITUATION des Schwedischen Feldlägers bey AMOENEBURG, wie auch die Einnehmung selbigen ohrts, durch Ihr Exellentz Herrn FeldMarschaln Carol Gustau Wrangel, den 19. Iuny Anno 1646.

Kay: Reduten

Käyserisch Jäger.

Schw: uorwacht

Moras

Kay: Reduten

Rideckon

Morus

Nider Ofelden

Mühl

Schweinberg

P

Mühl

S

Ohm: Au:

T

Omenburg

Omburg

Weg nach

Gundershausn

O

N

Erforthausen

F

I

Martorff

I

H

I

F

Roßtorff

Nider selheim

M

C

H

A

C

M

Holßhausen

Weg nach Gißen

Warttürn

G. W. Klenistratl. fecit

FORTRESS OF ATCHEH, SUMATRA, BY JOÃO TEIXEIRA ALBERNAZ Atcheh (Achin, Aceh), at the north-west tip of Sumatra, became a sultanate in about 1496, and, under Sultan Ali Mughayat Syah, conquered northern Sumatra by 1524, the year in which he defeated a Portuguese fleet. Thereafter, it was a centre of Islamic resistance to first the Portuguese and then the Dutch, although not always successfully. Both the Ottomans and the Islamic Sultanate of Golconda in India sent arms and men to help.

In 1629, the dynamic Sultan Iskandar Muda (r. 1607–36) was heavily defeated when, with his large galley fleet, he attacked the Portuguese-controlled port fortress of Malacca, in the Strait of Malacca, a funnel for trade between Malaya and Sumatra and thus between East and South Asia, and a port already unsuccessfully attacked by Atcheh in 1553, 1568, 1570, 1573 and 1595. However, Iskandar Muda was well up to the task of dominating northern Sumatra. **RIGHT**

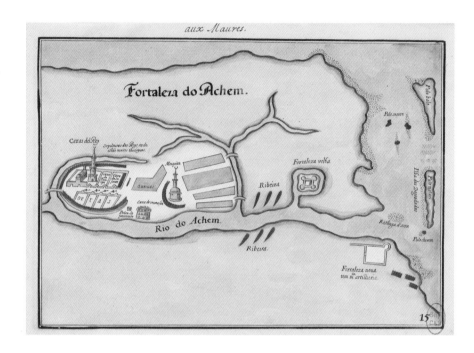

operations in foreign countries. Moreover, the National Land Survey of Sweden, which began in 1628, mapped not only Sweden at a large scale, but also lands occupied and annexed by the Swedes, for example Pomerania (coastal north-east modern Germany and coastal north-west modern Poland), and thereby provided a vital tool in planning defences as well as demonstrating success.

Similarly, the Russians made increased use of mapping during the century, especially toward the close. Prince Golitsyn, the lover of Peter the Great's half sister, Sophia (Regent from 1682 to 1689), made use of a primitive wagon-mounted odometer and compiled a *versta* or distance-book that was intended to be used to prepare a map for his march on the Crimean Khanate in the unsuccessful campaign of 1689 as part of Russia's war with the Ottoman (Turkish) Empire. His failure helped lead to Sophia's fall from power in favour of Peter (r. 1689–1725).

INCREASED PRECISION

More generally, the cartographic information available increased alongside expectations of precision. Whereas, for example, there had been considerable stylization in the depiction of physical features in medieval and early Renaissance maps (as the map-makers were primarily concerned with recording the existence of these features, rather than their accurate shape), the trend thereafter was towards precision in the portrayal of the crucial physical outlines, notably coastlines and rivers. This reflected the nature of the available information, as well as shifts in conceptual standards. At the same time, maps also moved away from pictorial (specific presentation) towards symbols (generalised representation).

Precision increased in a number of respects. For example, triangulation had been used to construct maps since the sixteenth century, but it was becoming more common. In 1679–83, the French *Académie* had worked out the longitudinal position in France. An improved ability to calculate longitude, combined with the use of triangulation surveying, affected mapping, both obliging and permitting the drawing of new maps. The direction of change in Europe was clear, while these skills were also taken elsewhere. Thus, Jesuits played a key role in the triangulation-based mapping of China, just as they also did in the introduction of improved artillery.

MILITARY ADAPTATION RATHER THAN REVOLUTION

From the mid-1950s, the period 1560–1660 was referred to by British scholars as one of 'military revolution', but this account no longer appears convincing. In many respects, instead, the seventeenth century saw a continuation of sixteenth-century developments. More specifically, the emphasis of the 'military revolution' thesis, as expanded by subsequent exponents, on firepower and *trace italienne* fortifications failed to capture, first, the extent to which the key theme was adaptation, rather than revolutionary change, as well as the diversity of developments, including the continued importance of cavalry.

The latter had an important implication for maps. The spatial dynamics of cavalry, the ground that could be covered in a period of time, as well as the impact on movement of terrain and cover, were very different to those for infantry. That prefigured later contrasts between infantry and armour and ensured that maps had to be read and understood in light of these very different requirements and possibilities.

As another central criticism of the 'military revolution' thesis, key changes occurred after 1660, rather than before it. This was true, in particular, of the use of the bayonet, which, by removing the need for pikemen, led to a marked increase in infantry firepower. So also did the introduction of the flintlock musket in place of the matchlock. The net effect was a change in infantry formations, one designed to maximise firepower in the context of differing defence requirements. In place of squares on a chequerboard pattern, infantry was increasingly deployed in linear formations.

Moreover, these formations became thinner and longer with time. This permitted more infantry to fire at once. Concentrated volley fire reduced the problems posed by low individual accuracy. The problems created by short-range muskets, which had a low rate of fire, and, as a consequence of their recoil, had to be re-sighted for each individual shot, were exacerbated by the cumulative impact of poor sights and poorly-cast shot. Battalions were drawn up generally only three ranks deep, and firings were by groups of platoons in a process designed to maximise the continuity of fire as well as the extent of fire control.

As a result of the changes, the battlefield spread out, which posed new issues of command and for 'situational awareness'. Both these encouraged a use of maps and diagrams. However, maps did, and do, not capture the constraints posed by the nature of the weaponry. This situation has only changed with recent maps on screen that permit the integration of dynamic elements and a variety of scales.

Since all the Christian European armies followed a similar pattern of re-equipment, capability gaps between them, and certainly marked capability gaps, did not emerge. In contrast, the weaponry, which offered greater firepower whatever its limitations, proved very useful against non-Westerners who lacked such weaponry. Much of the rest of the world did not follow the Christian European pattern of development. Thus, China did not move from matchlocks to flintlocks. Moreover, hand-to-hand weapons remained important there and elsewhere. The marked diversity of warfare needs to be remembered whenever the period and its mapping is considered.

SIEGE OF OSTEND. BY FRANCIS VERE (1560–1609).

PUBLISHED BY WILLIAM DILLINGHAM, 1657 The effective new Spanish commander, Ambrogio Spinola, captured Ostend in 1604 after a siege that lasted three years and seventy-seven days. Artillery bombardment was important, but so also was the willingness to storm positions, such as the outworks on the west side on 14 April 1603 and on Sand Hill on 13 September 1604, the last the decisive episode of the siege. The Dutch had kept Ostend resupplied from the sea, but the Spaniards forced its surrender by gaining control of the coastal sand dunes, which enabled them to mount batteries to dominate the harbour entrance. Spinola forced the pace of the siege by his willingness to sacrifice troops in determined assaults. Maurice of Nassau was unwilling to risk battle in order to relieve Ostend and preferred to besiege Sluys. As so often, operations in the field (either a battle or the decisions, as in this case, not to engage) helped determine the fate of a fort, as the garrison now had no hope of relief. **RIGHT**

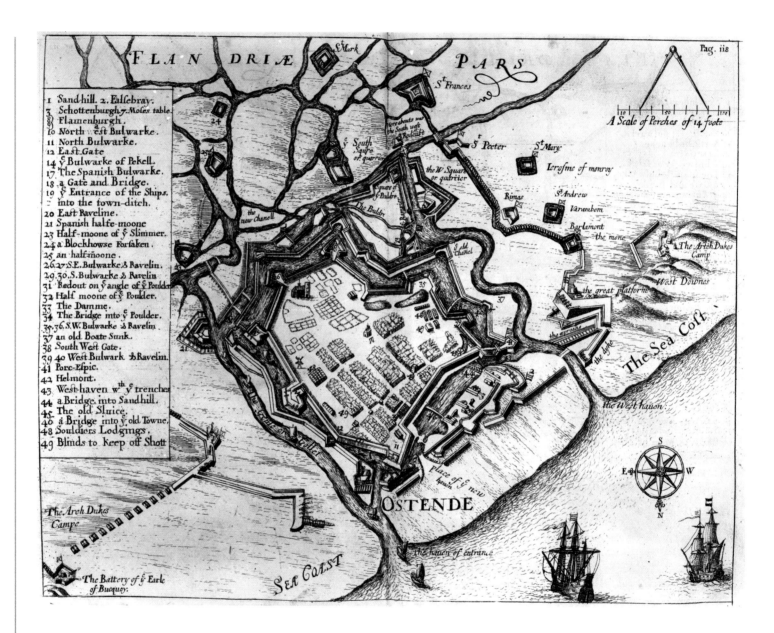

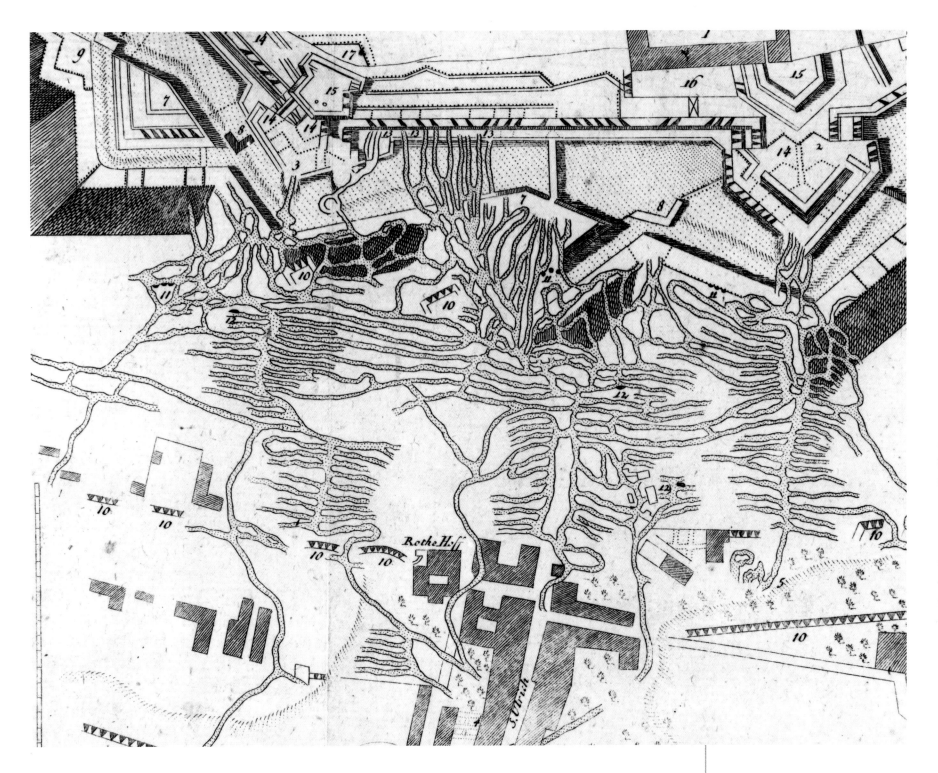

SIEGEWORKS DURING THE OTTOMAN SIEGE OF VIENNA, 1683 The main Ottoman force had left Adrianople (Edirne) on 31 March 1683, reaching Belgrade on 3 May and surrounding Vienna on 16 July. This date, however, was already fairly late in the year for a successful campaign, not least as the defences, ably improved by Georg Rimpler, were not suited for a general assault, but required a siege. With a relief force gathering, the Ottomans began building siegeworks, using both bombardment and mines to weaken the defences. The mines were particularly threatening and were designed to create breaches that would prepare the ground for assaults. The garrison suffered heavy casualties in its defence as well as losses from dysentery. The Ottomans, who were poorly prepared for a siege of such a powerful position with a very deep moat and large ramparts, suffered similarly. However, during August, the city's outer defences, where not covered by water features, steadily succumbed. Lacking heavy-calibre cannon, the outgunned Ottomans relied on undermining the defences, which they did with some success, leading to breaches in which there was bitter fighting. On 4 September, the garrison fired distress rockets to urge the relief army to action. It was to do so on 12 September, inflicting a serious defeat. ABOVE

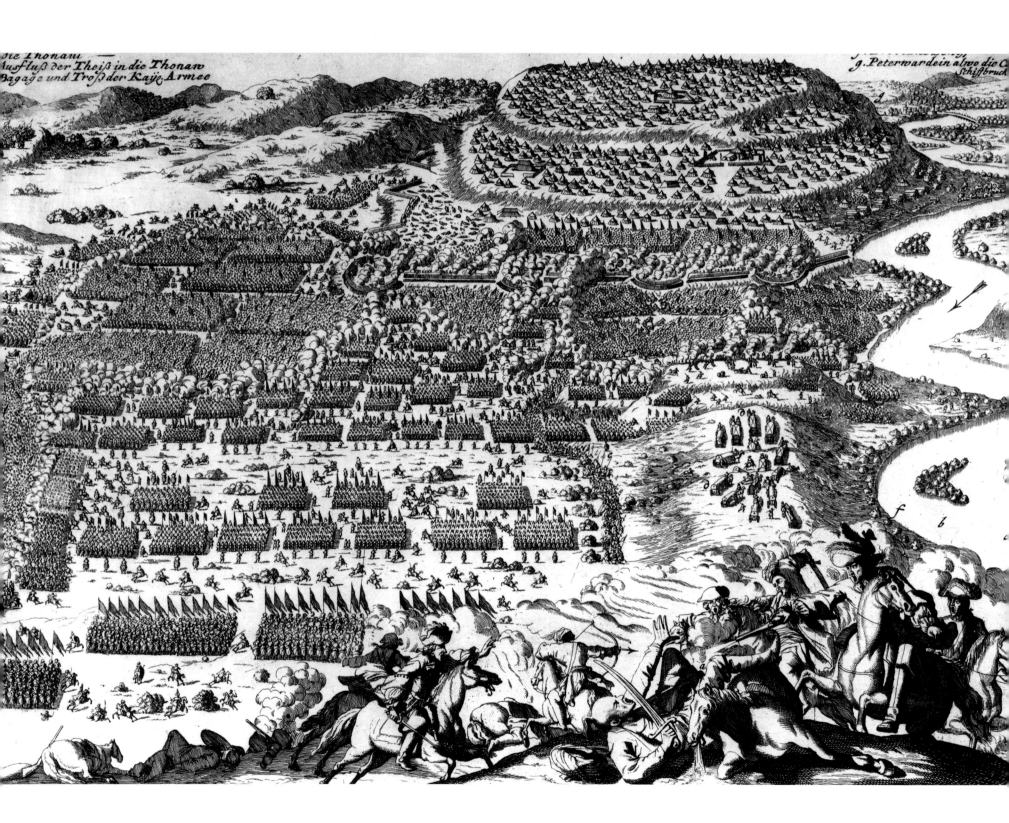

BATTLE OF ZÁDÁKEMÉN, 1691 In the 1680s, the Austrian advance into the Balkans had led to bold hopes about Turkish collapse. However, in 1690, under the able new Grand Vizier Fazil Mustafa, the Turks mounted a counter-offensive, recapturing the towns of Nish and Belgrade. This ensured that the Austrians moved most of their troops from facing the French to their eastern front. Three-quarters of the Austrian army served there in 1691. In 1691, Fazil Mustafa's hopes of recapturing Hungary were dashed by Prince Ludwig Wilhelm of Baden, a veteran of the battle of Vienna (1683), at Zádákemén, a long, hard-fought battle that left a third of the Imperial army killed or wounded, but the Turks routed, Fazil Mustafa killed and all the Turkish cannon captured. FAR LEFT

SIEGE OF PONDICHERRY, 1693 Established by François Martin as a French base on the Coromandel coast of India in 1674, the position was fortified in 1682, but was captured by the Dutch in 1693, bringing the French East India Company to the verge of bankruptcy. Pondicherry was regained by the French under the Treaty of Ryswick in 1697, and developed in the early eighteenth century. It had a population of about 120,000 by 1741 and was well fortified, a rampart with bastion on the side facing the land approaches being constructed in 1724–35, and a rampart on the coast in 1745. However, Pondicherry was captured by the British in 1761. LEFT

MAPS OF WAR

BREISACH. FRENCH SCHOOL Breisach was held by France from 1638, when it was starved into surrender by a pro-French force under Bernhard, Duke of Saxe-Weimar, until 1697, when it was returned under the Treaty of Ryswick. This was an important fortress in the Rhine valley that protected a crossing point, covered Alsace from invasion and opened the way into Germany. It was also important to Spanish routes between northern Italy and the Low Countries. The major and expensive fortifications programme under Louis XIV saw Vauban work on New Breisach in order to lessen the significance of the control of Breisach. The French also had to cede the fortresses of Kehl, Freiburg, Philippsburg and Luxembourg in 1697. **RIGHT**

Alsace

le village ruiné
des Bissen ancien quartier
du Ringraue de l'année
1634

Le Rhin de Bissen

l'hostellerie du
mortier

la redoute

taverne s. Jaques

le Pont

L'Isle s. Iaques

Projet du
fort de l'isle
s. Iaques

le moulin des Italiens

le Pont

Le Rhin

Porte du Rhin

ville basse

Kalterberg

la montagne dite Eysemberg

LE BRISGAW

Ce projet monstre de combien les boulewarts ont esté
rehaussés sauoir le Boulewart de Richelieu, ceux du souxtoch
et de vveymar auec la moittié de celluy de Kalterberg, selon quoy
seront parachevez les autres boulewart l'annee qui vient sauoir
l'allemand, le francois et le suedois.

FORTRESS ON THE BOSPHORUS. SEVENTEENTH CENTURY The stylised representation in this miniature from an Arabic manuscript of the seventeenth century captures the heavily fortified nature of Turkish positions, but also the conventional form they took, rather than the lower, earth (not stone)-reliant *trace italienne* form of new-style Western fortifications. RIGHT

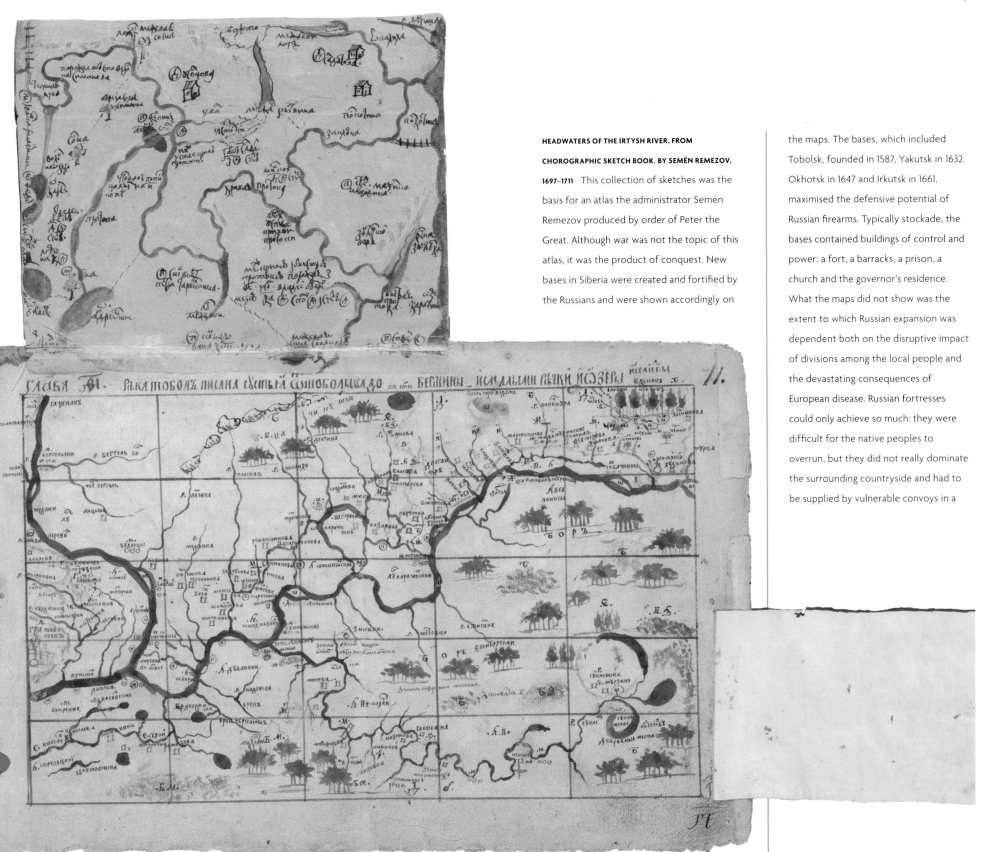

HEADWATERS OF THE IRTYSH RIVER, FROM CHOROGRAPHIC SKETCH BOOK. BY SEMËN REMEZOV, 1697–1711 This collection of sketches was the basis for an atlas the administrator Semën Remezov produced by order of Peter the Great. Although war was not the topic of this atlas, it was the product of conquest. New bases in Siberia were created and fortified by the Russians and were shown accordingly on the maps. The bases, which included Tobolsk, founded in 1587, Yakutsk in 1632, Okhotsk in 1647 and Irkutsk in 1661, maximised the defensive potential of Russian firearms. Typically stockade, the bases contained buildings of control and power: a fort, a barracks, a prison, a church and the governor's residence. What the maps did not show was the extent to which Russian expansion was dependent both on the disruptive impact of divisions among the local people and the devastating consequences of European disease. Russian fortresses could only achieve so much: they were difficult for the native peoples to overrun, but they did not really dominate the surrounding countryside and had to be supplied by vulnerable convoys in a

Professional MILITARY PLANNING

BATTLE OF RAMILLIES, 1706 On a spread-out battlefield, in a six-hour battle of roughly equal armies on 23 May 1706, John Churchill, 1st Duke of Marlborough showed the characteristic features of his generalship. Cool and composed under fire, brave to the point of rashness, he was a master of the shape and the details of conflict. He kept control of his forces and of the flow of the battle, and proved able to move and commit his troops decisively at the most appropriate moment. On the pattern of his 1704 triumph at Blenheim, Marlborough obtained a victory over Marshal Villeroi by breaking the French centre after it had been weakened in order to support action on the flanks. The French lost all their cannon and had about 19,000 casualties, compared with 3,600 casualties in Marlborough's army. The battle was followed by the rapid fall to Marlborough of many positions in the Low Countries, including Antwerp, Ath, Bruges, Brussels, Menin and Ostend. Villeroi, who had been more successful in blocking Marlborough in 1703 and 1705, held no further commands. RIGHT

WATERCOLOUR SKETCH OF A BATTLE DURING THE FOURTH AUSTRO-TURKISH WAR, 1737–1739. Contemporary Austrian painting. PREVIOUS PAGES

MAPS BECAME very much more significant in diplomacy and warfare with the eighteenth century as part of the development both of a new form of military professionalism and of the public discussion of politics. Encouraging the map culture, there was a general downgrading of theory in favour of facts, not least as the notion of applied knowledge acquired definition and prestige, especially as received wisdom was increasingly superseded by new information. Linked to this, there was a more explicit process of planning. Scientific methods entailed not only the concern of generals with artillery and sieges, but also the use of scientific knowledge at the operational level, with the need to plan foraging and marches requiring detailed information. Logistical skills were important to the staff planning that was at an increased premium.

The establishment of accurate values for longitude, one of the major practical and organisational achievements of the century, ensured that it became possible to locate most places accurately, and the development of accurate and standard means of measuring distances made it easier for map-makers to understand, assess and reconcile the work of their predecessors. Maps also became more predictable as mapping conventions developed. Even at the end of the seventeenth century, there was no standard alignment of maps, which was an aspect of a more general lack of standardisation and, even, precision. However, in the following century, the convention of placing North at the top was established.

An increased awareness of cartographic distinctiveness and change encouraged the idea that maps could, and thereby should, improve and respond to new and more accurate information. The correct relative location of features was expected, as was correct proportionality and keeping to scale. In addition, aside from specific improvements in mapping

techniques and concepts, maps were increasingly created for general reference, a process that enhanced their use.

The military were the major source for mapping across much of Europe. The longstanding surveying and charting facilities and interests of Western armies and navies varied greatly, but proved very important in the eighteenth century. In terms of scale, comprehensiveness and accuracy, the surveys then were in a different class to those a century earlier. Drawing on the cartographic traditions of their varied possessions, especially Italy and the Austrian Netherlands (modern Belgium), the Austrians were especially prominent in this mapping, in their far-flung possessions of Sicily, Lombardy, Austria, Bohemia, Hungary and the Austrian Netherlands. Ruling Sicily between 1720 and 1734, the Austrians employed army engineers to prepare the first detailed map of the island. As an aspect of Austrian defence preparations against Prussia, a major military survey of Bohemia (the western, major, part of the modern Czech Republic, already surveyed in 1712–18) was begun in the 1760s and completed in the 1780s: the Prussians had invaded Bohemia in the 1740s, 1750s and 1770s and were to do so again in 1866. In turn, Frederick II (the Great, r. 1740–86) of Prussia had his conquest of Silesia mapped.

The resulting detail was often considerable. The maps of the Austrian Netherlands drawn up in 1771–74 under the supervision of Joseph Johann de Ferraris, the director-general of the artillery, were based on military surveys and comprised 275 leaves. At this stage, there was no equivalent in Britain and, had the Austrians remained in control of the region instead of losing it to French invaders in 1792, then this mapping might have established a tradition that enjoyed the prestige of the later Ordnance Survey.

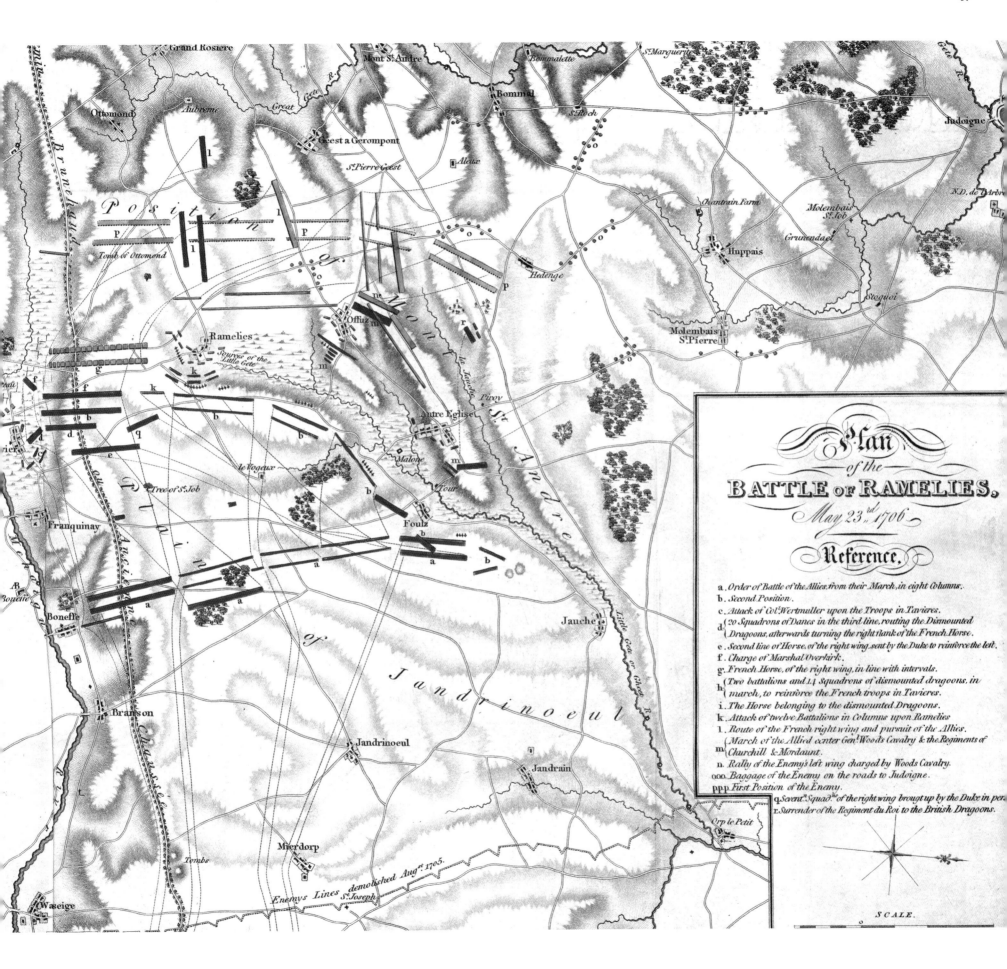

Plan
of the
BATTLE of RAMELIES,
May 23rd 1706

Reference.

a. Order of Battle of the Allies, from their March, in eight Columns.
b. Second Position.
c. Attack of Colᵗ Wertmuller upon the Troops in Tavieres.
d. 20 Squadrons of Danes in the third line, routing the Dismounted Dragoons, afterwards turning the right flank of the French Horse.
e. Second line of Horse, of the right wing, sent by the Duke to reinforce the left.
f. Charge of Marshal Overkirk.
g. French Horse, of the right wing, in line with intervals.
h. Two battalions and 14 Squadrons of dismounted dragoons, in march, to reinforce the French troops in Tavieres.
i. The Horse belonging to the dismounted Dragoons.
k. Attack of twelve Battalions in Columns upon Ramelies.
l. Route of the French right wing and pursuit of the Allies.
m. March of the Allied center Genᵗ Woods Cavalry & the Regiments of Churchill & Mordaunt.
n. Rally of the Enemy's left wing charged by Woods Cavalry.
ooo. Baggage of the Enemy on the roads to Judoigne.
ppp. First Position of the Enemy.
q. Sevenᵗʰ. Squadᵗⁿˢ of the right wing brougt up by the Duke in per...
r. Surrender of the Regiment du Roi to the British Dragoons.

SCALE.

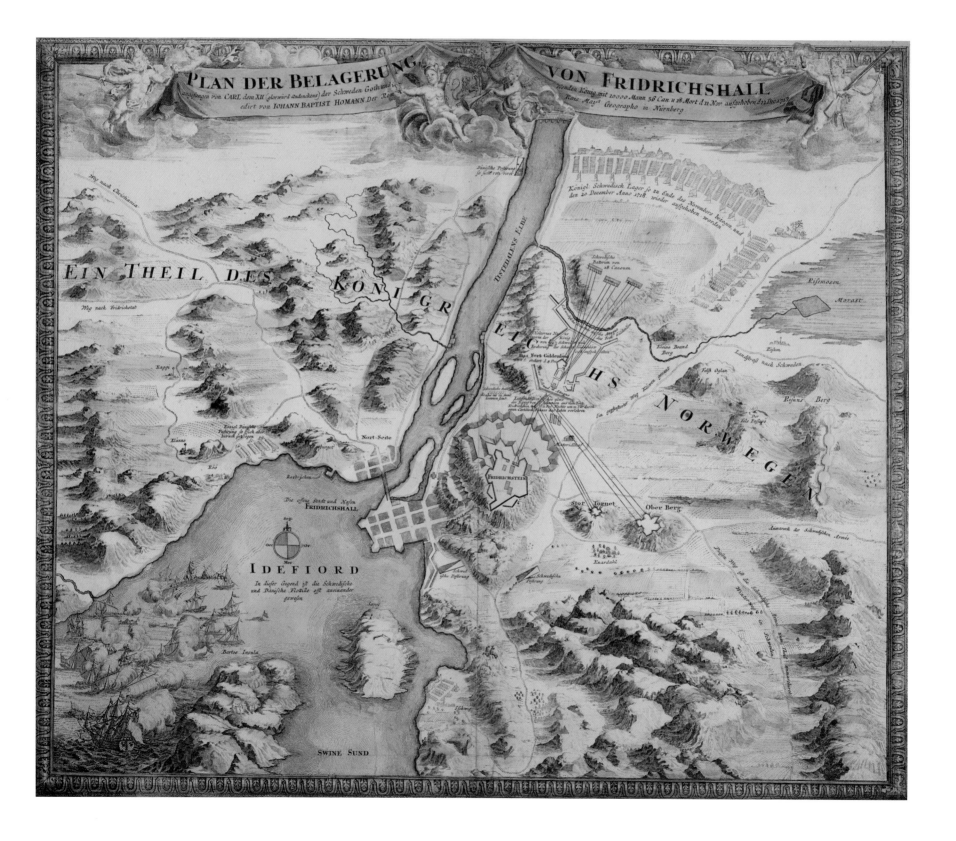

Military engineers from other countries were also important. Following the suppression of the 1745 Jacobite rebellion with William, Duke of Cumberland's victory over 'Bonnie Prince Charlie', Charles Edward Stuart, at Culloden in 1746, William Roy of the British Royal Engineers between 1747 and 1755 carried out a military survey of Scotland that served as the basis for an accurate map. Six surveying parties were employed. The map was produced at a scale of one inch to 1000 yards. Survey and maps were designed to help in the British military response to any future rebellion, and to assist in the process of the governmental reorganisation of the Highlands. This was a key aspect of a longer term extension of central control that included road building and the establishment of fortified positions, notably the still-impressive Fort George near Inverness.

NEW HEIGHTS

Moreover, the French military engineers of the period, notably Pierre-Joseph Bourcet, tackled the problems of mapping mountains, creating, as a result, a clearer idea of what the Alpine frontier looked like. His work on the Alps was continued by Jean Le Michaud d'Arçon. The mapping helped the French Revolutionary forces in their successful campaigns there in the 1790s, notably those of Napoleon, although this mapping does not explain why the French were more successful than in the 1740s when they had last contested this area. Bourcet also mapped Corsica, as part of the brutal imposition of French control there in the face of a bitter rebellion in 1768–70 after the island was purchased from Genoa in 1768. Success in overcoming this rebellion, a process demonstrating the far greater resources of the French monarchy, ensured that Napoleon was born a French subject in 1769.

In Piedmont, the *Ufficio degli Ingegneri Topografi* was founded in 1738, with four topographical engineers recruited. The hiring of assistants ensured continuity and, from 1743, all the personnel were employed on the fronts as Piedmont was at war with France and Spain in the War of the Austrian Succession. Piedmontese mapping tended to be confidential and the maps remained in manuscript.

In Denmark, more widespread use and production of cartographic material to be employed at the tactical scale began to appear during the century, and notably from the 1770s.

War repeatedly led to an increased demand for geographical information, at a number of levels. Thus, the silk maps printed for military use in northern Italy at the Milanese press of Marc' Antonio Dal Re in the 1730s and 1740s, during the Wars of the Polish (1733–35) and Austrian (1740–48) Successions, for example *Italiae septentrionalis* (1735) and *Nuova carta corografica, o sia centro del gran teatro di guerra in Piemonte Savoia l'anno 1744* (1744), were very useful for route planning. The war in northern Italy could also be followed in Johannes Covens and Cornelis Mortier's *Le cours du Po*, published in Amsterdam in 1735. In the *Pensées sur la tactique, et la stratégique* (1778), Marquis Emanuele de Silva Taroicca, an officer in the army of Victor Amadeus III of Sardinia, ruler of Savoy-Piedmont pointed out that, because maps could be defective, generals needed to gain personal knowledge of the terrain.

ACROSS OCEANS

From the 1750s, there was also a greater emphasis on trans-oceanic campaigning. This became more important for Britain and France from 1754 as far as North America was concerned. There the operational theatre spanned hundreds of miles. In a way that is well-nigh inconceivable today, British and French

SIEGE OF FREDERIKSTEN FORTRESS. BY JOHANN BAPTISTE HOMANN, 1718 A key episode in the Great Northern War (1700–21), where the Swedish army laid siege to this fortress, near Frederikshald, on the border between Sweden and Danish-ruled Norway. The Swedish trenches had almost reached the main fortification walls on the night of 11 December 1718 when a bullet struck and killed Charles XII of Sweden while he inspected the trenches. His death effectively ended the Swedish invasion plans. LEFT

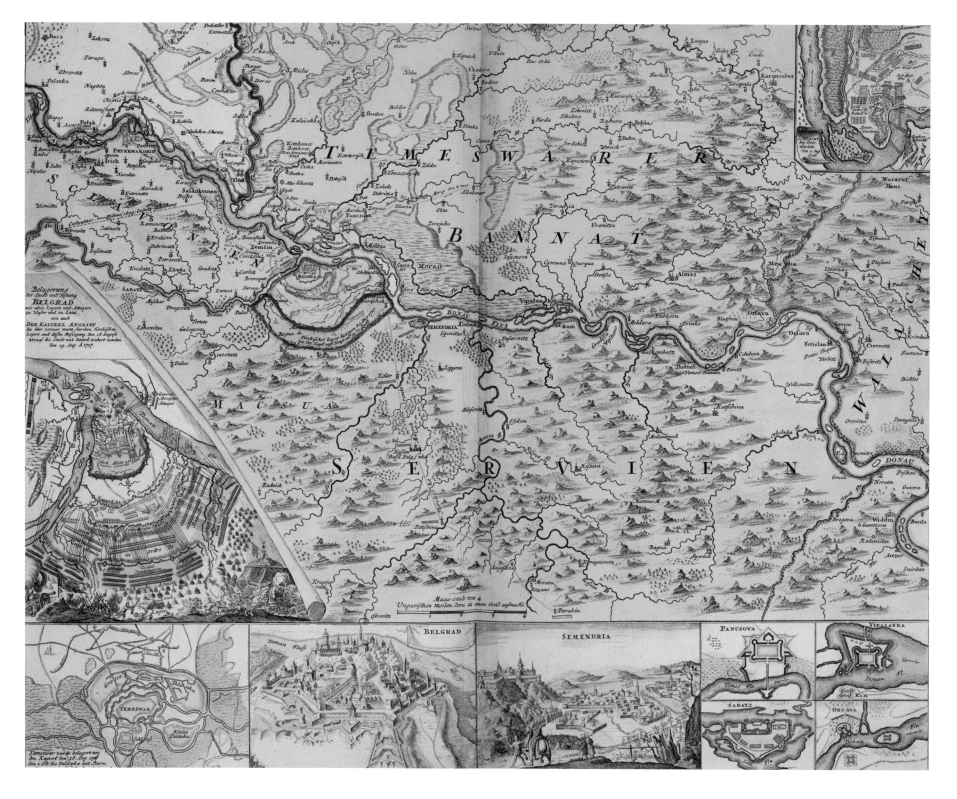

CELEBRATING AUSTRIAN VICTORY. BY JOHANN BAPTIST HOMANN, 1720 Soon after the 1716–18 war between Austria and the Turks ended, Johann Baptist Homann, a leading Nuremberg map-maker, produced a map providing, alongside the theatre of war, insets showing the principal battles and fortresses involved. On 5 August 1716, under Prince Eugene, the Austrians smashed the Turks at Petrovaradin (Peterwardein), an Austrian base on the right bank of the River Danube north-west of Belgrade. The success of the Austrian cavalry proved decisive in this battle. Eugene then marched on the mighty fortress of Temesvár, which had successfully defied the Austrians in the 1690s and which controlled or threatened much of eastern Hungary. On 16 October, Temesvár surrendered after heavy bombardment. In June 1717, advancing anew, Eugene crossed the Danube by pontoon bridges east of Belgrade, which was surrounded and besieged. A Turkish relief army arrived and the Austrians attacked and beat the latter on 16 August in a confused engagement fought in the fog. The garrison of Belgrade, now denied the chance of relief, surrendered. The campaigns of 1716 and 1717 suggested that, however well fortified, the fate of fortresses was likely to be settled by battle, as defeat denied the possibility of relief and therefore exposed defenders to the prospect of unremitting supply problems. In addition, defeat in battle entailed a crucial loss of prestige, which also encouraged surrender.

ABOVE

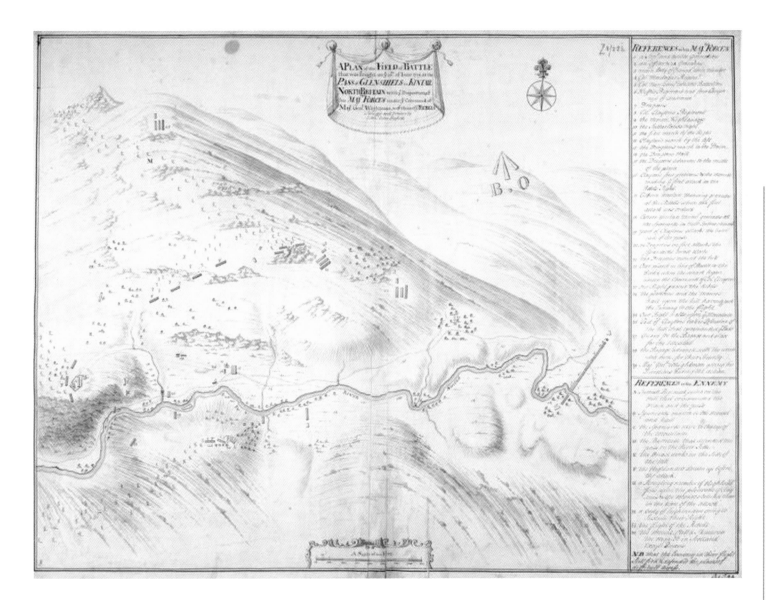

BATTLE OF GLENSHIEL. BY JEAN HENRI BASTIDE,
1719 The problems of depicting terrain
emerge clearly from this map of the
Battle of Glenshiel; at the same time, the
detailed key showed what could be
covered. In 1719, two Spanish frigates
carrying 300 Spanish troops under
George Keith, the Earl Marischal, a
Jacobite, reached Scotland. It was
intended to tie down British troops in
Scotland, while the main expedition
invaded England, but the latter was
dispersed by a violent storm. This
discouraged much of the Highlands from
supporting the Earl Marischal, while the
government forces acted rapidly and
decisively. The two sides met in Glenshiel
on 10 June. The Jacobite-Spanish force,
1,850 strong, remained in a good
defensive position that gave them the
potential to mount a charge, but
Major-General Wightman, with 1,100
government troops, took the initiative
and, assisted by mortar fire, successfully
attacked the Jacobite flanks. The Jacobite
army disintegrated, with the Spaniards
surrendering and the Highlanders retiring
to their homes. LEFT

forces had to traverse vast distances, often through wild, inhospitable terrain, making large armies vulnerable to enemy tactics of *petite guerre*, by which small, highly mobile detachments carried out fleeting attacks and ambushes on the flanks of their larger adversary. Commanders therefore required maps that both gave an account of routes and terrain suitable for planning and conveyed very practical knowledge that would allow their forces to move quickly without being over-exposed to enemy action.

An absence of maps could have serious consequences, both in North America and elsewhere. For example, it hindered the Dutch in their unsuccessful operations into the interior of Sri Lanka (Ceylon) in 1764. The kingdom of Kandy had long resisted the Europeans settled on the coast, first the

Portuguese and then the Dutch. Because of the problems encountered in 1764, the Dutch Governor Lubbert Jan, Baron Van Eck had new maps drawn and, in 1765, the Dutch launched a new campaign. They replaced swords and bayonets with less cumbersome machetes, provided troops with a more practical uniform and moved more rapidly, which helped them gain the initiative and lessened the impact of disease. To begin with, the Dutch triumphed, capturing the deserted capital, Kandy, to which they were guided by their new maps. However, the Kandyans refused to engage in battle, and Dutch energies were dissipated in seeking (and failing) to control a country rendered intractable by disease and enemy raiders. Peace was made in 1766 and Kandy would not be conquered until the British overran it in 1815. Thus, maps were

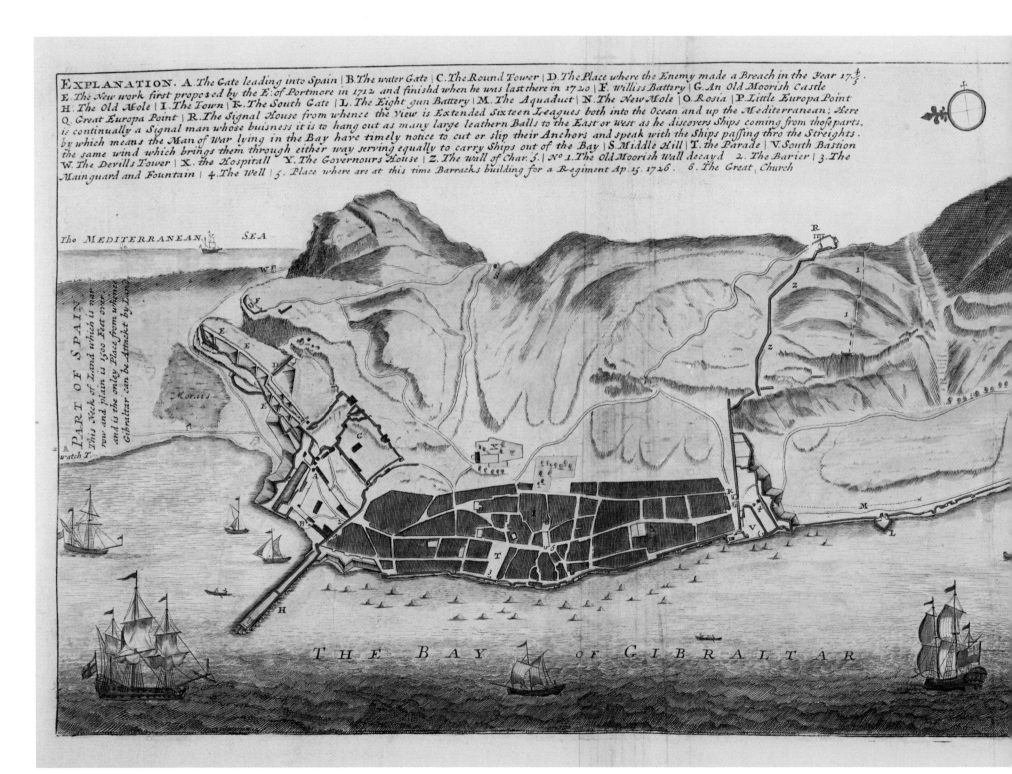

EXPLANATION. A. The Gate leading into Spain | B. The water Gate | C. The Round Tower | D. The Place where the Enemy made a Breach in the Year 17.4/5.
E. The New work first proposed by the E: of Portmore in 1712 and finishd when he was last there in 1720 | F. williss Battery | G. An Old Moorish Castle
H. The Old Mole | I. The Town | K. The South Gate | L. The Eight gun Battery | M. The Aquaduct | N. The New Mole | O. Rosia | P. Little Europa Point
Q. Great Europa Point | R. The Signal House from whence the View is Extended Sixteen Leagues both into the Ocean and up the Mediterranean; Here
is continually a Signal man whose buisness it is to hang out as many large leathern Balls to the East or West as he discovers Ships coming from those parts,
by which means the Man of War lying in the Bay have timely notice to cut or slip their Anchors and speak with the Ships passing thro the Streights.
the same wind which brings them through either way serving equally to carry Ships out of the Bay | S. Middle Hill | T. the Parade | V. South Bastion
W. The Devills Tower | X. the Hospitall | Y. The Governours House | Z. The wall of Char. 5. | No 1. The Old Moorish Wall decayd   2. The Barier | 3. The
Mainguard and Fountain | 4. The Well | 5. Place where are at this time Barracks building for a Regiment Ap. 15. 1726.   6. The Great Church

The MEDITERRANEAN SEA

PART OF SPAIN
This Neck of Land which is nar
row and plain is 1500 Feet over
and is the onley Place from whence
Gibraltar can be Attackt by Land

Morais

THE   BAY   OF   GIBRALTAR

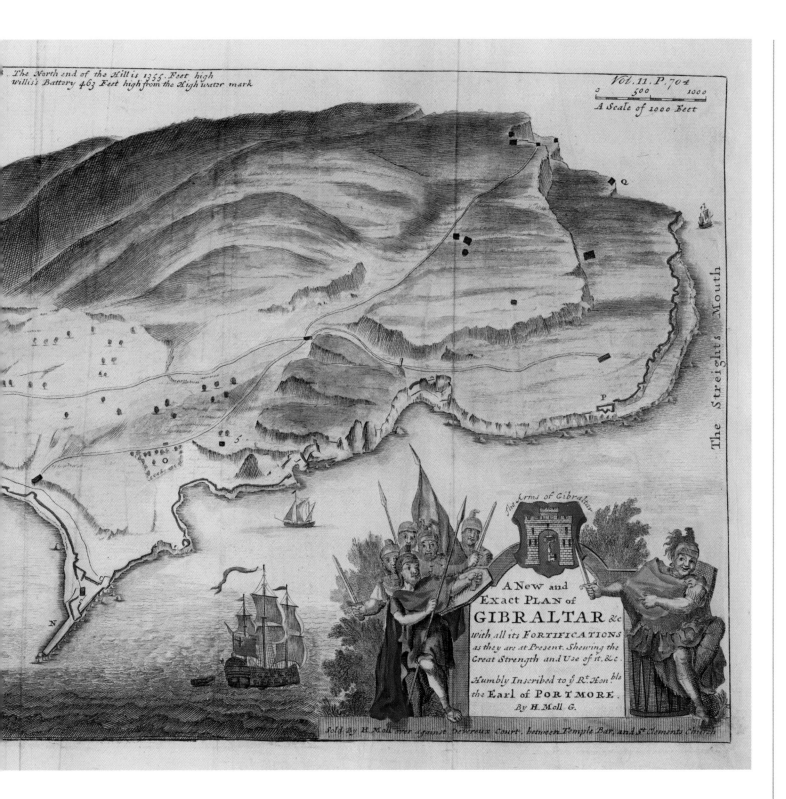

The North end of the Hill is 1355 Feet high
Willis's Battery 463 Feet high from the High water mark

Vol. II. P. 704

A Scale of 1000 Feet

The Streights Mouth

The Arms of Gibraltar

A New and
Exact PLAN of
GIBRALTAR &c
with all its FORTIFICATIONS
as they are at Present, Shewing the
Great Strength and Use of it. &c.

Humbly Inscribed to y.e R.t Hon.ble
the Earl of PORTMORE.
By H. Moll G.

Sold by H. Moll over against Devreux Court, between Temple Bar, and St Clements Church

GIBRALTAR FORTRESS UNDER SIEGE, BY HERMANN MOLL, 1727 Hermann Moll produced his *New and Exact Plan of Gibraltar* in 1727 because the British fortress was besieged that year by the Spaniards. Gibraltar had fallen to a British amphibious attack in 1704. It had then been defended, largely by naval deployments and operations, especially in 1704 and 1705, but subsequently by an impressive series of fortifications. The combination of the two elements ensured that Gibraltar did not fall during a series of conflicts between Britain and Spain. Restricted largely to a blockade and a light bombardment, the 1727 siege was not pressed home, unlike that in 1779–83, a major and protracted siege that led to the appearance of more plans. **LEFT**

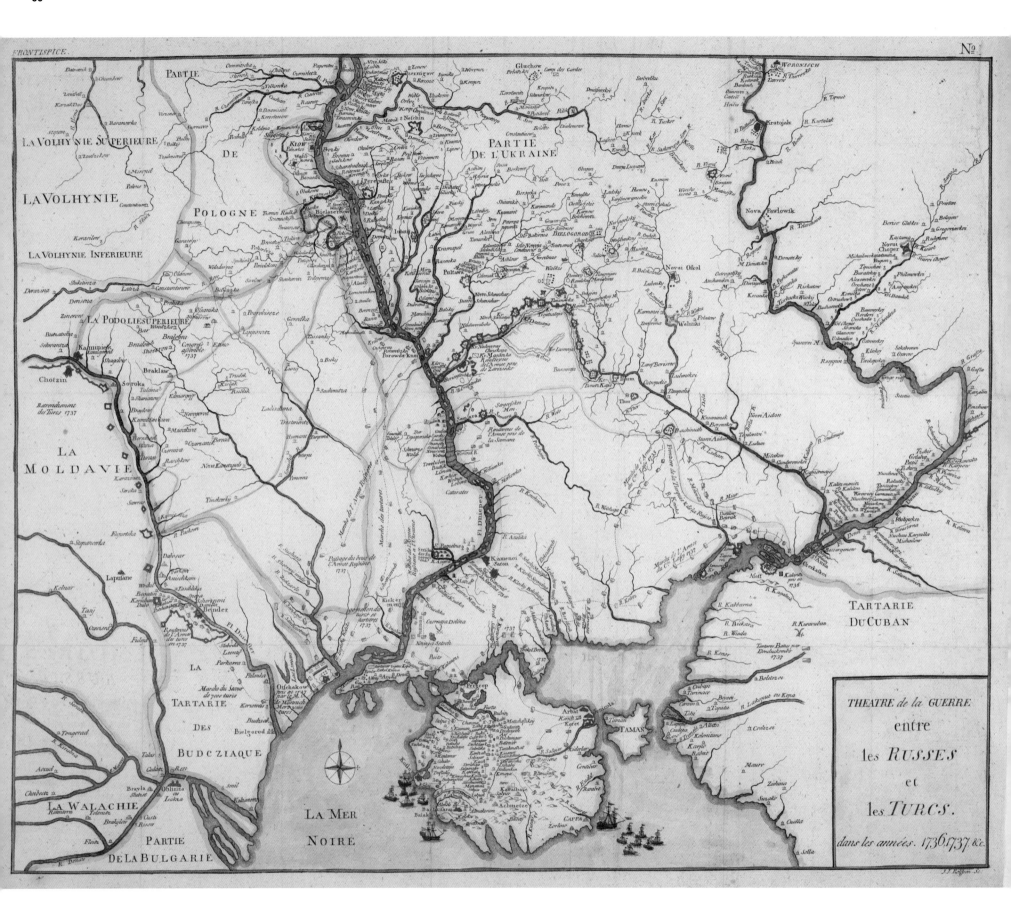

THEATRE de la GUERRE
entre
les RUSSES
et
les TURCS.
dans les années 1736, 1737. &c.

only significant as part of a broader range of capability. The nuances of this situation were different in operations between European powers.

In North America, European commanders ordered the preparation of reconnaissance maps and route maps. The former involved the rapid production of information, and the resulting maps were generally crude. For example, in February 1775, just months before the Battle of Bunker Hill, General Thomas Gage, the British commander, dispatched two amateur surveyors to reconnoitre the countryside around Boston, which was the key British garrison point and the centre of political contention.

ROUTE MAPS

Route maps, which were of more particular significance, were usually executed by professional military surveyors. While they may have incorporated intelligence gleaned from reconnaissance maps, these maps laid out specific routes for the army to take, often also drawing on information from (sometimes already existing) topographical surveys conducted under scientific conditions. This mapping was frequently undertaken with specific reference to an intended itinerary, and considered daily progress and the most appropriate sites for encampments and resupply of the force. As a result, an accurate and formal detailing of distances and features of the landscape was crucial.

Perhaps the finest set of route maps produced during the century were those by the Comte de Rochambeau's engineers that painstakingly charted the route of the French army from their base at Newport, Rhode Island, to attack the British army at Yorktown, Virginia, in 1781. This carefully-chosen route over a significant distance afforded the army the speed and stealth that allowed it to arrive at its objective at an ideally opportune moment, while

avoiding British detection. As a result, George Washington was able to bring overwhelming force to bear against the British under Charles, Earl Cornwallis, at Yorktown. Whether or not accompanied by maps (or by maps that have survived), the planning of itineraries was a key element of military activity.

STRATEGY AND PLANNING

Maps played a role in strategic discussion and planning. This was seen clearly in 1712, during the negotiations over ending the War of the Spanish Succession when Torcy, the French foreign minister, urged his British counterpart, Bolingbroke, to look at a map in order to see the strategic threat posed by the Alpine demands of Victor Amadeus II of Savoy-Piedmont.

In his novel *Humphry Clinker* (1771), the journalist Tobias Smollett made the need for cartographic information apparent. This provided a way to mock Thomas, Duke of Newcastle, a leading British minister from 1724 to 1756 and 1757 to 1762, with reference to the position at the beginning of the Seven Years' War (1756–63):

This poor half-witted creature told me, in a great fright, that thirty thousand French had marched from Acadie [Nova Scotia] to Cape Breton – 'Where did they find transports? (said I)' 'Transports! (cried he) I tell you they marched by land' – 'By land to the island of Cape Breton?' 'What! Is Cape Breton an island?' 'Certainly.' 'Ha! Are you sure of that?' When I pointed it out in the map, he examined it earnestly with this spectacles; then, taking me in his arms, 'My dear C---! (cried he) You always bring us good news – Egad! I'll go directly, and tell the king that Cape Breton is an island.'

In practice, there is no information on how far Newcastle read maps. Indeed, such information is rare

THE THEATRE OF WAR IN UKRAINE DURING 1736–37, BY J. J. ROLFFSEN, C. 1780 In 1736, Russia declared war on the Turks, seized Azov, after the main powder magazine blew up, and invaded Crimea, storming the earthworks that barred the isthmus of Perekop at the entrance to Crimea in a night attack in columns. The Russians then invaded Crimea, occupying the towns and burning down the palace of the khans. However, the Russians suffered from a Tatar scorched earth policy in which crops were burnt and wells poisoned. Disease and heat were also major problems. The following year, the Russians, under General Burkhard von Münnich, took the fortress of Ochakov on the estuary of the River Bug. They were hindered by the Tatars who burnt the grass between the Rivers Bug and Dniester. Logistical issues and disease also caused problems. **LEFT**

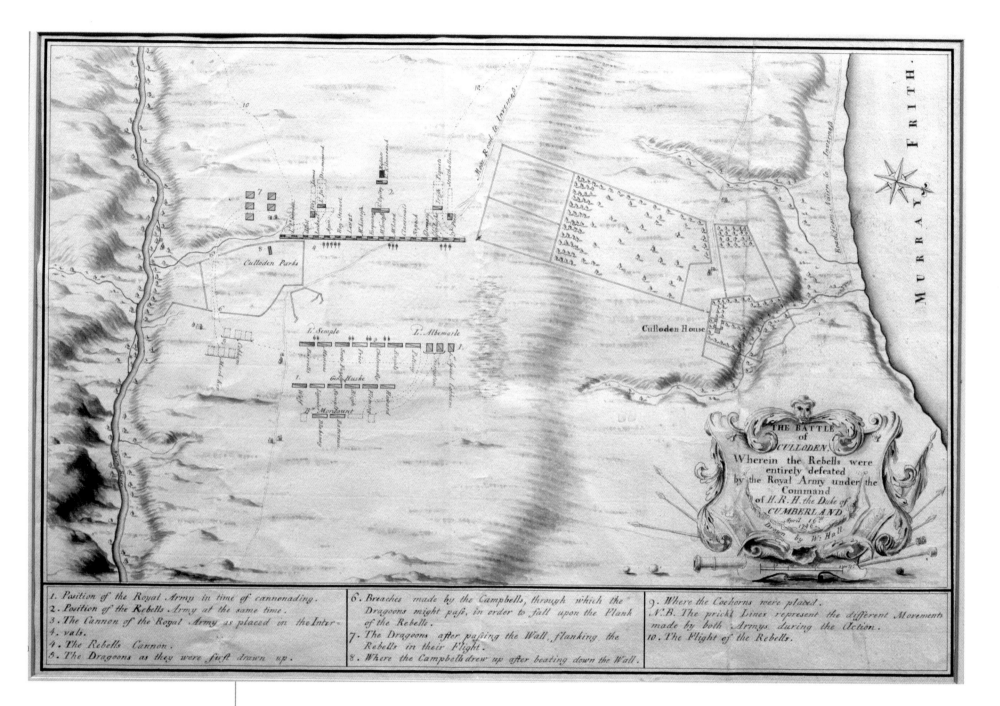

MURRAY FRITH.

Culloden Parks

L. Semple   L. Albemarle

Gen. Huske

B. Mordaunt

Culloden House

THE BATTLE
of
CULLODEN,
Wherein the Rebells were
entirely defeated
by the Royal Army under the
Command
of H.R.H. the Duke of
CUMBERLAND
April 16th 1746
Drawn by W. Hall

1. Position of the Royal Army in time of cannonading.
2. Position of the Rebells Army at the same time.
3. The Cannon of the Royal Army as placed in the Inter-
4. vals.
4. The Rebells Cannon.
5. The Dragoons as they were first drawn up.

6. Breaches made by the Campbells, through which the
Dragoons might pass, in order to fall upon the Flank
of the Rebells.
7. The Dragoons after passing the Wall, flanking the
Rebells in their Flight.
8. Where the Campbells drew up after beating down the Wall.

9. Where the Coehorns were placed.
N.B. The prickt Lines represent the different Movements
made by both Armys during the Action.
10. The Flight of the Rebells.

BATTLE OF CULLODEN, 16 APRIL 1746. BY WILLIAM HALL

This plan focuses on the location of units, rather than the dynamics of the battle which were set by the Highland charge and its inability to break Cumberland's force. Firepower so thinned the numbers of the advancing Jacobite clansmen that their chance of defeating their more numerous opponents was limited. The general rate of fire was increased by the level ground and the absence of any disruptive fire from the Jacobites. Many factors led to confusion among the Jacobites: the slant of their line, the nature of the terrain which was partly waterlogged, the difficulty of seeing what was happening in the smoke produced by the guns, and the independent nature of each unit's advance. Cumberland wrote: 'In their rage that they could not make any impression upon the battalions, they threw stones at them for at least a minute or two, before their total rout began.' ABOVE

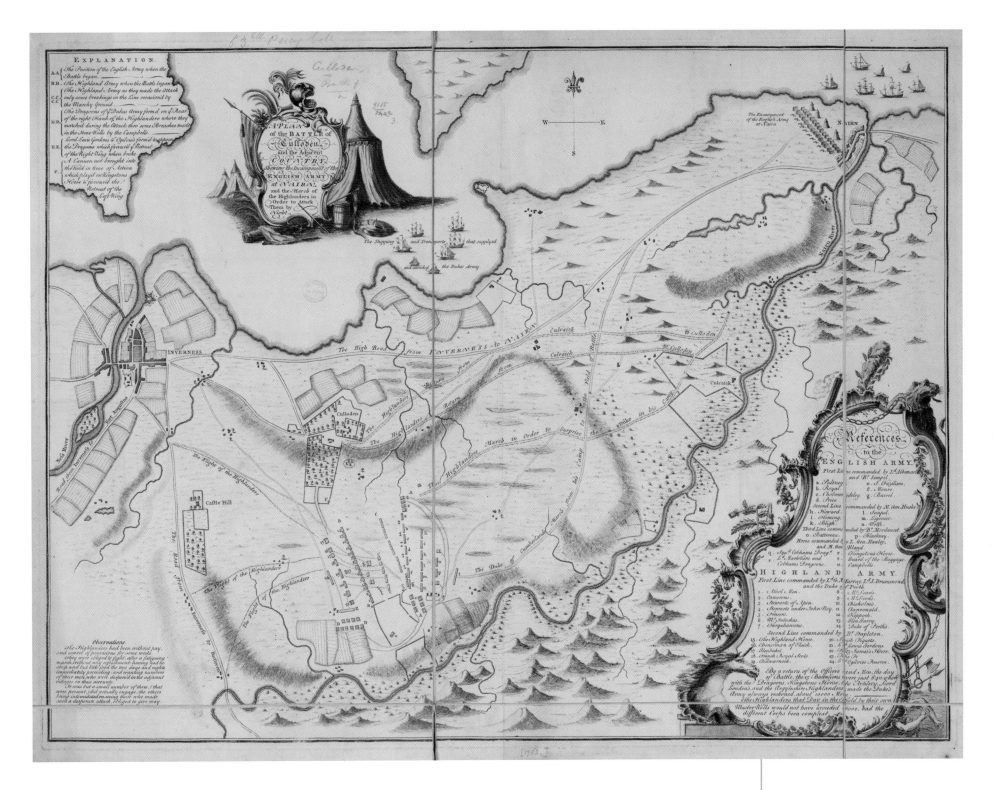

CULLODEN AND THE SURROUNDING COUNTRYSIDE, BY J. FINLAYSON, 1752 An unopposed fording of the River Spey near its mouth on the North Sea, on 12 April 1746, enabled the Duke of Cumberland to advance westward on Inverness. His men were well paid and fed and their morale was high, elements that could not be represented on a map. The same goes for the problems with the morale of the Jacobite army whose food supply had collapsed. On 14 April, Cumberland reached Nairn. Charles Edward tried a night attack on Cumberland's force on the night of 15–16 April, but the poorly-conceived plan was abandoned due to delays. Both sides readied for battle on the 16th. ABOVE

BRITISH CAMP ON CUBA, 1741 The British invasion of Cuba was a failure for reasons that this map of the well-defended position they established in Guantanamo Bay did not reveal. An attack on the major port of Santiago was planned, but the British troops were landed in the bay more than eighty miles away. This foolishly exposed them to a long and dangerous advance through woody terrain ideal for Spanish guerrilla action. The troops suffered heavily from disease, did not reach their goal and were re-embarked. Santiago was to fall to American attack in 1898. An earlier British attack on Cartagena (in modern Colombia) and a later one on Panama, both in 1741, also failed. RIGHT

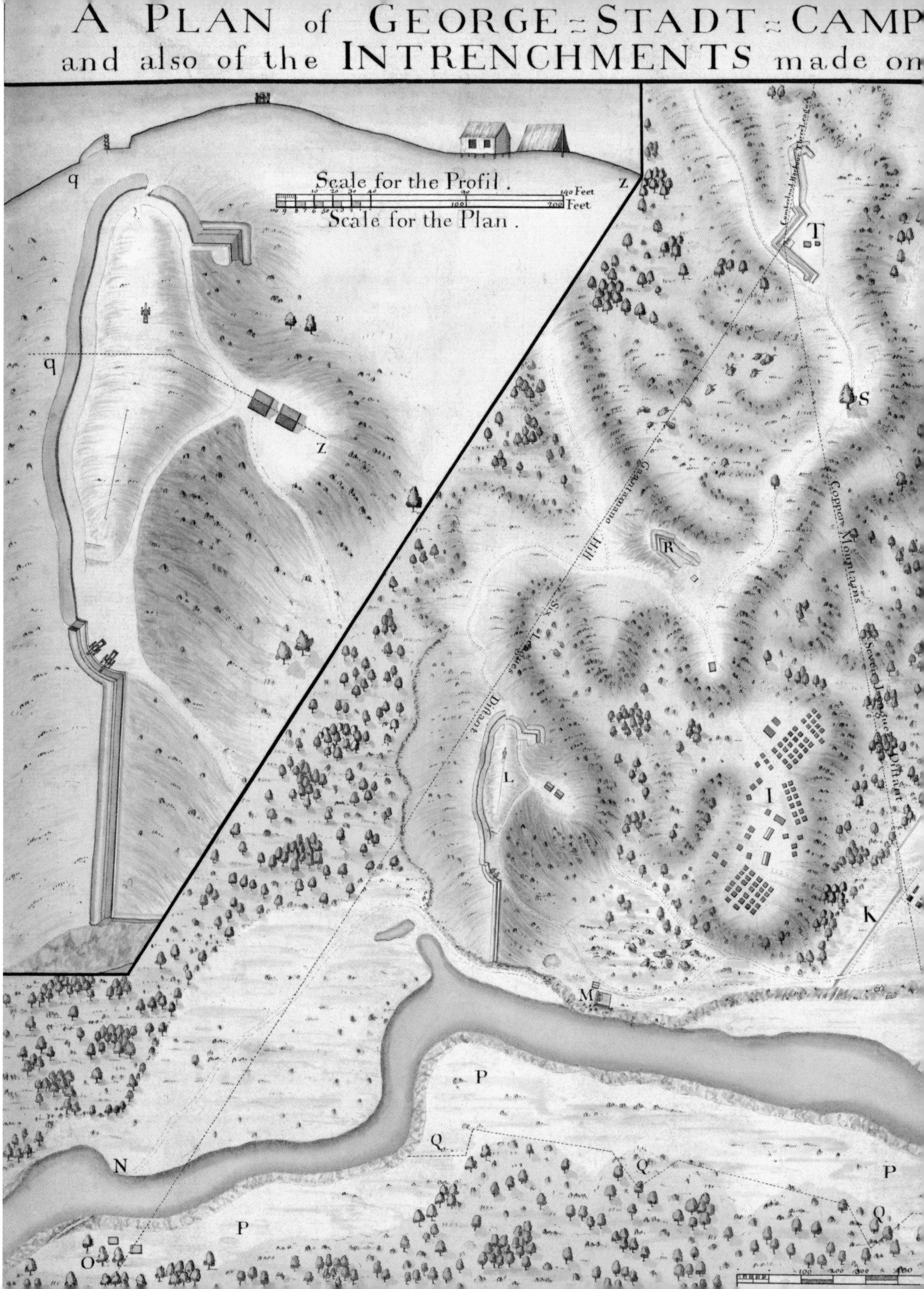

A PLAN of GEORGE-STADT-CAMP
and also of the INTRENCHMENTS made on
Scale for the Profil.
Scale for the Plan.
140 Feet
700 Feet
Guantanamo Hill
Six Leagues Distant
Copper Mountains
Seven League Distant
q
q
z
z
T
S
R
L
I
K
M
P
Q
N
O

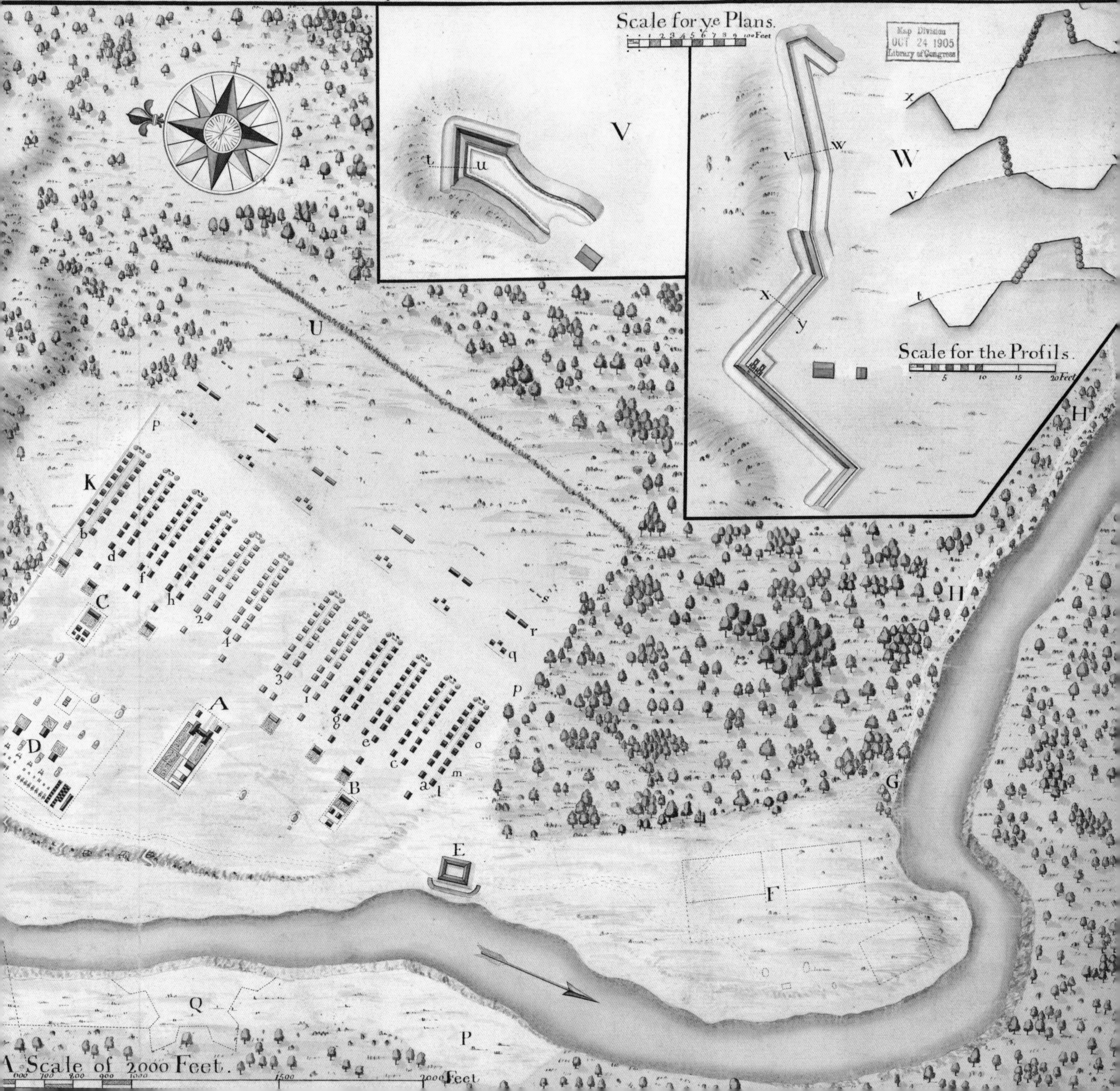

Scale for ye Plans.

1 2 3 4 5 6 7 8 9 100 Feet

V

W

Scale for the Profils.

5 10 15 20 Feet

U

H

K

P

b

d

C

f

h

2

4

A

3

i

g

D

B

e

c

a

m

o

q

r

s

p

G

E

F

H

Q

P

P

A Scale of 2000 Feet.

100 200 300 400 500 600 700 800 900 1000 1500 2000 Feet

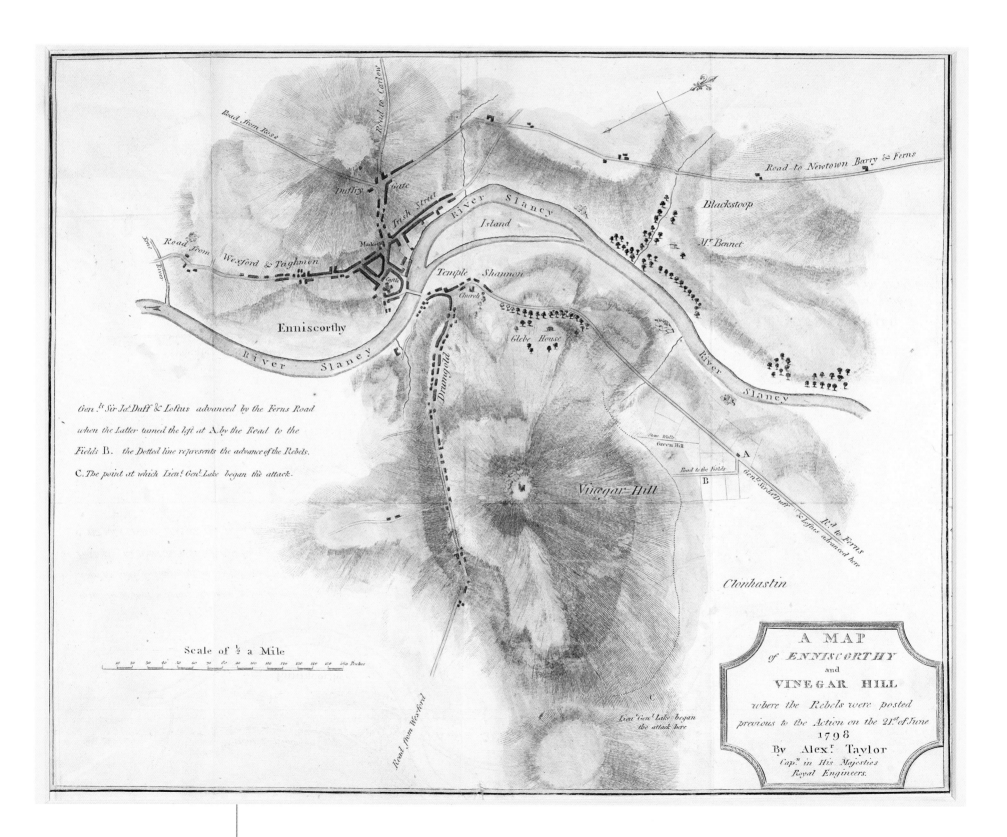

Road to Carlow

Road from Ross

Duffry Gate

Irish Street

River Slancy

Island

Road to Newtown Barry & Ferns

Blackstoop

Mr Bennet

Road from

Wexford & Taghmon

Tone River

Market

Castle

Temple Shannon

Church

Enniscorthy

River Slaney

Glebe House

River Slaney

Gen.ls Sir Ja. Duff & Loftus advanced by the Ferns Road

when the Latter turned the left at A. by the Road to the

Fields B. the Dotted line represents the advance of the Rebels.

C. The point at which Lieut. Genl. Lake began the attack.

Draumghid

Stone Walls

Green Hill

Road to the Fields

A

Gen.ls Sir Ja Duff & Loftus advanced her

Rd to Ferns

B

Vinegar-Hill

Clonhastin

Scale of ½ a Mile

10 20 30 40 50 60 70 80 90 100 110 120 130 140 150 160 Poles

Road from Wexford

C

Lieut Genl Lake began
the attack here

A MAP
of ENNISCORTHY
and
VINEGAR HILL
where the Rebels were posted
previous to the Action on the 21.st of June
1798
By Alexr. Taylor
Capn in His Majesties
Royal Engineers.

for this period. During the Dutch crisis of 1787, a map played a role in the advice offered the British Cabinet then led by William Pitt the Younger. The advice was offered in person by Charles, 3rd Duke of Richmond, the Master-General of the Ordnance (artillery), the sponsors of the Ordnance Survey, and Sir James Harris, the experienced British envoy in The Hague. Richmond 'talked of military operations – called for a map of Germany – traced the marches from Cassel and Hanover, to Holland, and also from Givet to Maastricht'. The former would be the route followed by troops from Hesse-Cassel and Hanover sent to help the Orangists, the latter anticipated that the French would intervene on the side of their Dutch protégés, the Patriots. The following day, Harris saw Pitt, who 'sent for a map of Holland; made me show him the situation of the [United] Provinces [the Netherlands]'.

A map would also demonstrate the contrast between the distance allied Prussian troops would have to travel from their Rhineland base in Cleves, in order to mount an invasion of the Netherlands, and the greater distance France's forces would have to take from Givet in order to intervene on behalf of the Patriots. This advice was designed to encourage British support for action on behalf of the Orangists by illustrating its viability. In the event, the British lent naval backing, the Prussians successfully invaded, and French preparations did not result in action. Thus senior figures made detailed reference to maps.

Indeed, in 1792, George III of Britain used a map to follow the Prussian invasion of France, an invasion finally checked by a larger French army at Valmy, an army that also benefited from stronger artillery.

PUBLIC INTEREST
Wars encouraged public interest in maps and entrepreneurial publishers responded. Thus, the

Huguenot (French Protestant) exile Abel Boyer published in London in 1701 *The draughts of the most remarkable fortified towns of Europe … With a geographical description of the said places. And the history of sieges they have sustain'd.* This was a moment in international tension that was propitious for entrepreneurs, as the wide-ranging and lengthy War of the Spanish Succession (1701–14, Britain involved 1702–13) was beginning. Four years later appeared in Paris a collected edition of Nicolas de Fer's fortification plans, *Les forces d'Europe*, which had first been published between 1690 and 1695, during the Nine Years' War.

In 1781, the 'advertisement' for General Henry Lloyd's *Continuation of the history of the late war in Germany, between the King of Prussia, and the Empress of Germany and her Allies* noted:

In order to elucidate in one view the particular reflections and descriptions contained in this work, as well as in military history in general, a map on a large scale is now engraving, that will comprehend the countries between the Meridian of Paris and that of Petersburg, and from the latitude of the last mentioned place, to that of Constantinople; on which will be traced the natural lines of operation, leading from the frontiers of the respective countries; as also the lines on which the respective armies did really act in the several campaigns during the war we describe, which will enable the reader to see and judge of the propriety of their operations. This map will be given to purchasers of the work.

The reasons why individual maps were published are not always so easy to gauge, but maps appearing in wartime presumably served an audience with interests accordingly. Thus, in 1708, Henry Pratt's wall map of Ireland was advertised in London as at sale for a guinea (£1.05p today). This was a map that would have

BATTLE OF VINEGAR HILL. BY J. HARDY, 1798 This plan of Vinegar Hill near Enniscorthy, County Wexford, shows the position of the armies on 21 June 1798. The key battle in the Irish Rising of 1798, this was a struggle won by the larger and better-armed side. The poorly-led rebels in Wexford concentrated at Vinegar Hill, losing the strategic initiative and allowing the British to land reinforcements near Waterford from 16 June. Lieutenant-General Gerard Lake was able to concentrate an army of 20,000 men and a large artillery train. He attacked his 9,000 opponents on 21 June, using his artillery to devastate them. The rebels fought for two hours, suffering heavy casualties, and finally retreated when their ammunition ran out. The rebel pikemen were shot down. Their cohesion lost, the rebels suffered heavily in subsequent government punitive operations. The rising had been defeated. **LEFT**

SIEGE OF MADRAS. MAURILLE ANTOINE MOITHEY,

1750 This coloured map depicts a French success of 1746 when 1,200 French regulars, supported by warships from the French colony of Mauritius, captured Fort St George after only a two-day siege. The garrison suffered only six dead, while the British fleet in the area refused to attack the larger French fleet. Madras was returned under the Peace of Aix-la-Chapelle of 1748, under which the British returned Louisbourg on Cape Breton Island. Madras did not fall again. RIGHT

helped ministers and civilians alike to follow any fighting that broke out there. Roads and barracks were depicted, while the map was flanked by town plans that made clear citadels and walls, for example at Kinsale. Moreover, that of Drogheda included a plan of the nearby Battle of the Boyne. Fought in 1690, this was the decisive engagement in which William III defeated James II.

In many cases, it was clear why specific maps appeared. In 1727, the second edition of Jacques Ozanam's *A treatise of fortification* was published in London with the addition of *A new and exact plan of Gibraltar with all its fortifications* by Hermann Moll, the latter plan made topical by the Spanish siege that year. Moreover, the publication of this edition as a whole was a response to the more general international crisis of 1725–27, one in which a large-scale war appeared very probable in 1726–27.

Publications also provided an opportunity for discussing fortification techniques, as with Bernard Forest de Bélidor's *La science des ingénieurs dans la conduit des travaux de fortification et d'architecture civile* (1729). Debates over military engineering, notably in France, indicated the controversial nature of best practice.

Sieges were understood in the context of maps of larger areas. Thus, the War of the Spanish Succession led to maps such as *A true & exact map of the seat of war in Brabant and Flanders with the enemies lines in their just dimensions* (1705). In this detailed map, the moves of Anglo-Dutch-German forces commanded by John, 1st Duke of Marlborough, and of his French opponents could be charted. This map was followed by the issue by John Harris of *A new map of Europe done from the most accurate observations communicated by the Royal Societies at London and Paris illustrated with plans and views of the battles, sieges and other advantages obtained by*

her Majesties Forces and those of Her Allies over the French. Insets in this map included plans of the battles of Blenheim (1704) and Ramillies (1706), both key victories for Marlborough.

Similar maps appeared for particular battles, enabling readers to understand the general context, as well as the course, of the battle. Thus, the battle of Falkirk in 1746, a major clash in the Jacobite rising that had started in 1745, was rapidly commemorated in a map showing the general area. It included a plan of the battle of Falkirk and of the battle at Culloden later that year. Culloden was also the subject of a separately published plan. In the same war, the British capture of the French base of Louisbourg on Cape Breton in 1745, the major French naval base in North America, had led to the publication of maps in Britain.

Battle and siege maps did not only appear in the immediate aftermath of conflict, but also subsequently. They were clearly seen as both interesting in themselves and as appropriate illustrative material. Thus, in the Marquis de Quincy's *Histoire militaire du règne de Louis le Grand* [Louis XIV], published in 1726, there were good battle maps. That of Ramillies (1706) included a key that provided an explanation of the battle, and thus a degree of dynamism to illuminate the plan of the battle.

NORTH AMERICA

War in North America in 1743–48, 1754–60 and 1775–82 led to the appearance of more maps of that continent, at least in so far as known by Europeans, and these were widely publicised. Thus, new maps of North America were announced in the issues of a London newspaper, the *Daily Advertiser*, on 3 August, 5 September, and 10 September 1755 as large-scale conflict between Britain and France began there in 1755.

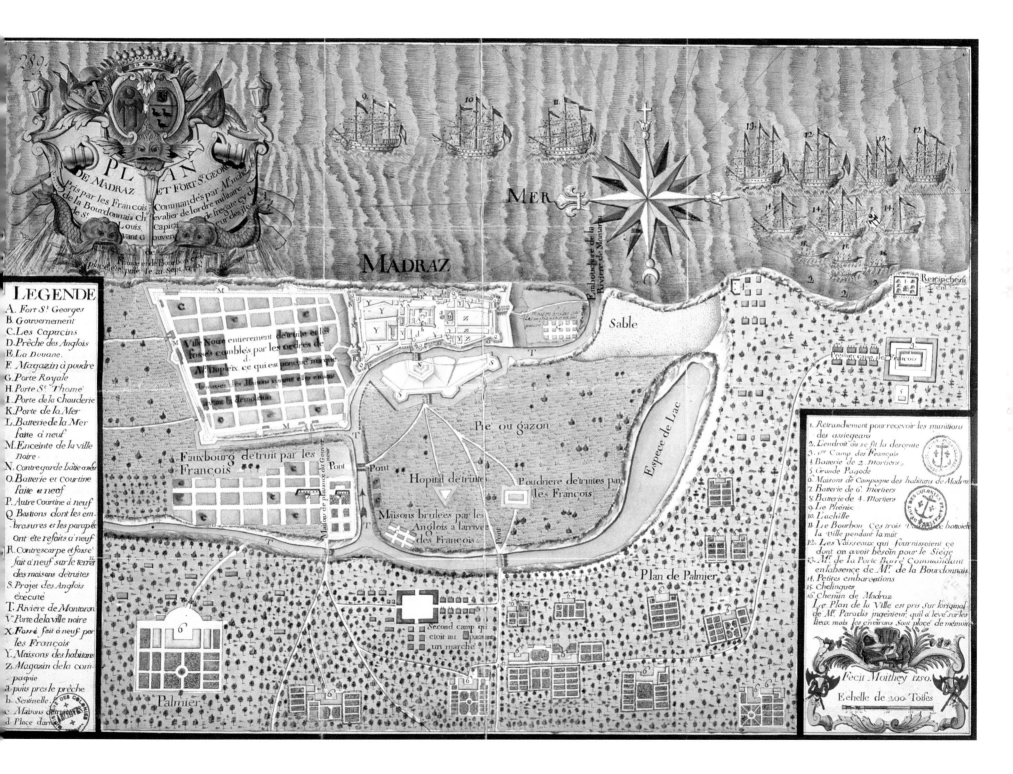

LEGENDE
A. Fort St Georges
B. Gouvernement
C. Les Capucins
D. Prêche des Anglois
E. La Douane
F. Magazin à poudre
G. Porte Royale
H. Porte St Thomé
I. Porte de la Chaudrerie
K. Porte de la Mer
L. Batterie de la Mer
   faite à neuf
M. Enceinte de la ville
   noire
N. Contregarde bâtie à neuf
O. Batterie et Courtine
   faite à neuf
P. Autre Courtine à neuf
Q. Bastions dont les em-
   brasures et les parapet
   ont été refaits à neuf
R. Contrescarpe et fossé
   fait à neuf sur le terrein
   des maisons détruites
S. Projet des Anglois
   exécuté
T. Riviere de Montaron
V. Porte de la ville noire
X. Fossé fait à neuf par
   les François
Y. Maisons des habitans
Z. Magazin de la com-
   pagnie
a. puits près le prêche
b. Sentinelle
c. Maisons détruites
d. Place d'armes

PLAN DE MADRAZ ET FORT St GEORGE
Pris par les François Commandés par Mr...
de la Bourdonnais Chevalier de l'ordre militaire...
...Louis, Capit...avant Gouvern...
...Place...prise le 21 Sept 1746

MER

MADRAZ

Embouchure de la Rivière de Montaron

Sable

Ville Noire entierement détruite et les
fossés comblés par les ordres de
Mr Dupleix ce qui est ponctué marque
les masses des Maisons comme elles étoient
avant la démolition

Fauxbourg détruit par les
François

Maison de plaisance du Gouverneur

Pont   Pont

Pie ou gazon

Hopital détruit

Poudriere détruite par
les François

Espece de Lac

Premier camp des François

Maisons brulées par les
Anglois à l'arrivée
des François

Pont

Palmier

Plan de Palmier

Second camp qui
étoit au paravant
un marché

Retranchement

1. Retranchement pour recevoir les munitions
   des assiegeans
2. L'endroit où se fit la descente
3. 1er Camp des François
4. Batterie de 2 Mortiers
5. Grande Pagode
6. Maisons de Campagne des habitans de Madras
7. Batterie de 6 Mortiers
8. Batterie de 4 Mortiers
9. Le Phénix
10. L'Achille
11. Le Bourbon Ces trois Vaisseaux battoient
    la Ville pendant la nuit
12. Les Vaisseaux qui fournissoient ce
    dont on avoit besoin pour le Siege
13. Mr de la Porte Barré Commandant
    en l'absence de Mr de la Bourdonnais
14. Petites embarcations
15. Chelingues
16. Chemin de Madraz
Le Plan de la Ville est pris sur l'original
de Mr Paradis ingenieur, qui à levé sur les
lieux mais les environs Sont placé de mémoire

Fecit Moithey 1750.

Echelle de 200 Toises

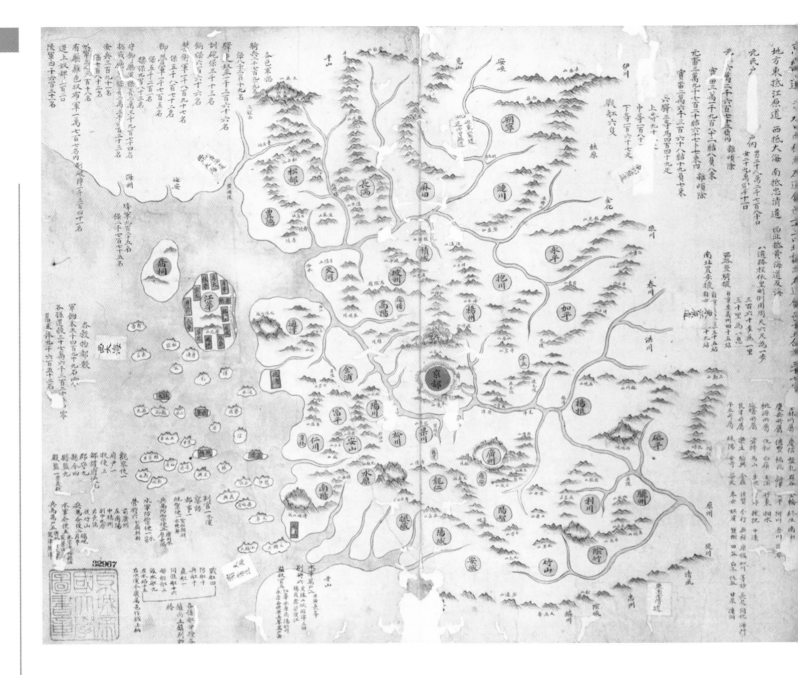

PROVINCE OF GYEONGGI-DO, BY THE GOVERNMENT OF KOREA, 1750s This map, produced by the government of Korea in the 1750s, shows the province of Gyeonggi-do, which literally means the royal capital and its vicinity. The red spot on the map marks Seoul. The map, which does not depict campaigning, can be compared with those of Europe and North America during this period in order to show the nature of the available maps that could be used for purposes of military planning. The map included key military bases and strategic points, including several fortresses built to defend Seoul. The map provides detailed geographical information judged crucial in governing the country. **RIGHT**

Maps also played a role in supporting the discussion of policy. John Clevland, Secretary to the Admiralty, had a copy of a series of tracts of 1755–56 that dealt with North America, tracts supported by maps. One of the tracts was Lewis Evans' *Geographical, historical, political, philosophical and mechanical essays; the first containing an analysis of a general map of the middle British colonies in America and of the country of the confederate Indians …* (London and Philadelphia, 1755). It was used by General Edward Braddock who advanced into the interior in 1755, only to be successfully ambushed by French and native forces near his intended destination, Fort Duquesne, near Pittsburgh.

The accuracy of maps was contested. Criticism of Evans' map led in 1756 to the publication of a second essay by Evans. In turn, Ellis Huske's *The present state of North-America* (1755) backed the cartographer John Mitchell in his pro-British *Map of the British and French dominions in North America* (1755). Public interest can be gauged from the fact that two editions of Huske appeared in Boston in 1755 and two in London. The accuracy of Mitchell's map was contested by Thomas Jefferys in his *Explanation for the new map of Nova Scotia … with the adjacent parts of New England and Canada* (1755).

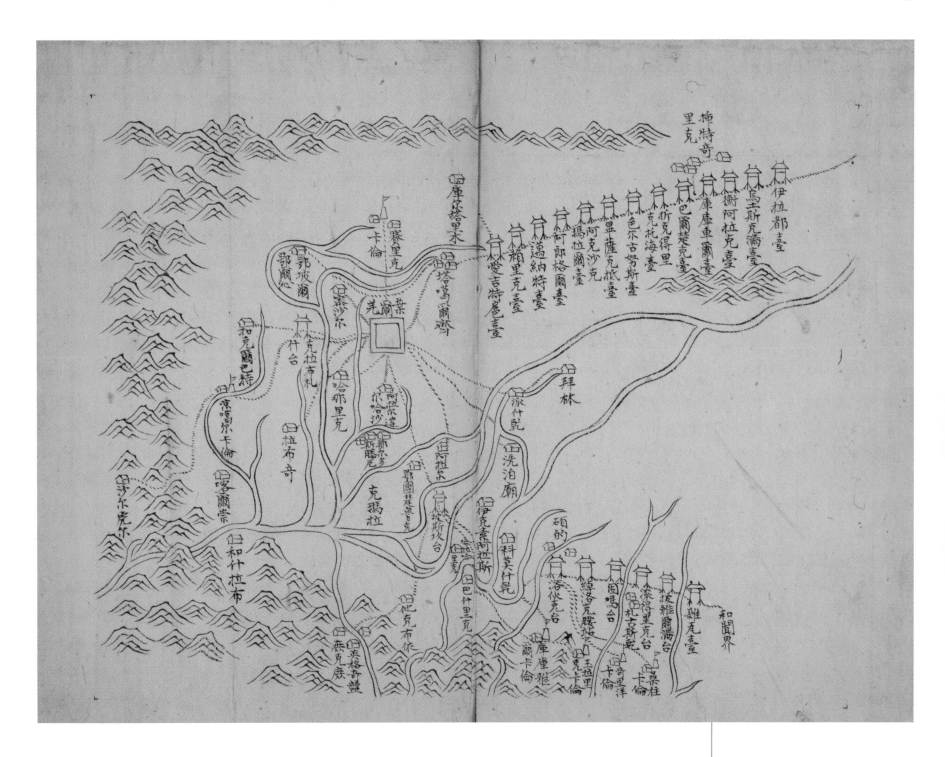

XINJIANG, 1759 This Chinese map shows the topography, cities, roads and military posts in Xinjiang. While not to scale, it ably captured the relationships between place and position and recorded the Chinese determination to protect the acquisitions gained in their successful advances in 1755 and 1757. In the 1750s, the Chinese established chains of magazine posts along the main roads. In 1755, two advances, each with 25,000 troops, were launched, the Northern Route army advancing into Xinjiang from the north-east via Outer Mongolia, and the West Route army, from Gansu, further south, which moved forward via Hami. After a rebellion in 1755–6, the Chinese advanced anew in 1757. In 1759, they pressed on to capture the towns of Kashgar and Yarkand in eastern Turkestan. ABOVE

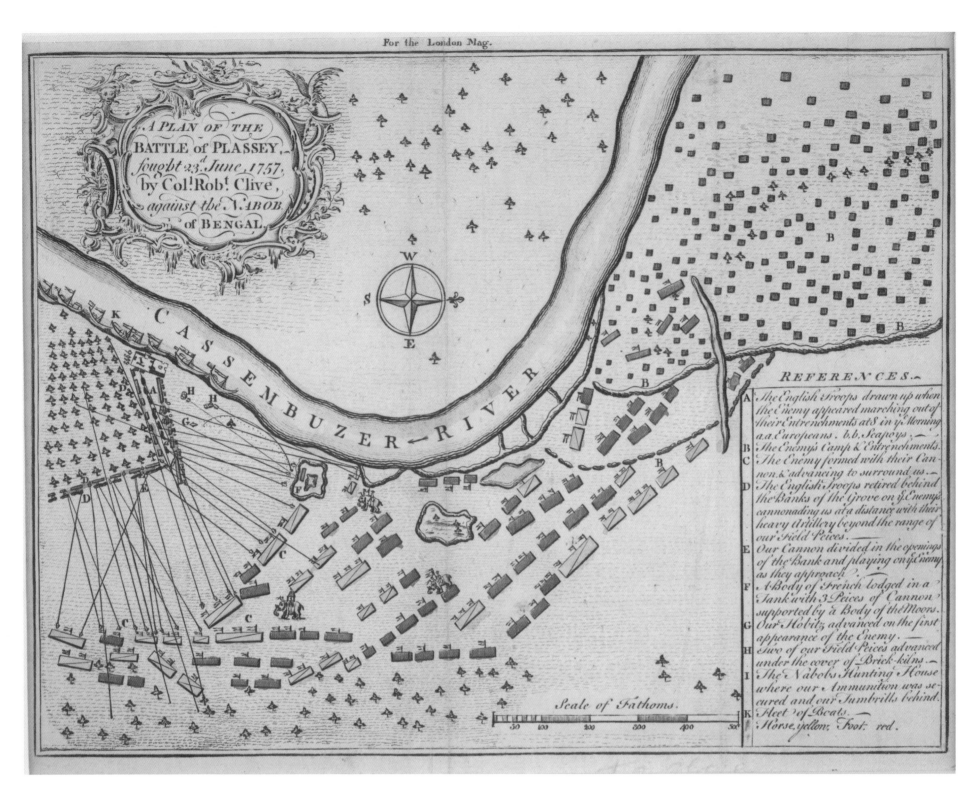

A PLAN OF THE
BATTLE of PLASSEY,
fought 23.d June 1757,
by Col.l Rob.t Clive,
against the NABOB
of BENGAL.

CASSEMBUZER RIVER

REFERENCES.

A The English Troops drawn up when
the Enemy appeared marching out of
their Entrenchments at 8 in y.e Morning
a.a. Europeans. b.b. Seapoys.
B The Enemy's Camp & Entrenchments.
C The Enemy formed with their Cannon & advancing to surround us.
D The English Troops retired behind
the Banks of the Grove on y.e Enemy's
cannonading us at a distance with their
heavy Artillery beyond the range of
our Field Peices.
E Our Cannon divided in the openings
of the Bank and playing on y.e Enemy
as they approach.
F A Body of French lodged in a
Tank with 3 Peices of Cannon
supported by a Body of the Moors.
G Our Hobitz advanced on the first
appearance of the Enemy.
H Two of our Field Peices advanced
under the cover of Brick-kilns.
I The Nabob's Hunting House
where our Ammunition was secured and our Tumbrills behind.
K Fleet of Boats.
Horse, yellow, Foot, red.

Scale of Fathoms.
50 100 200 300 400 50

BATTLE OF PLASSEY. MAP PUBLISHED IN 1760S

Advancing on the capital of Siraj ud-Daulah, the Nawab of Bengal, who had stormed Fort William at Calcutta the previous year, Robert Clive encountered the latter's army near the village of Plassey on 23 June 1757. Clive deployed his men, about 850 British troops and 2,100 sepoys (Indians), in front of a mango grove with an acute angle of the river behind him, the sepoys on the flanks and his ten field guns and howitzers in front; fifty sailors acted as artillerymen. An artillery duel began and Clive withdrew his men into the grove, where they sheltered behind the mud banks and among the trees. The Indians, about 50,000 strong, made no real effort to attack the British position, with the exception of a cavalry advance that was driven back by grapeshot. A torrential midday downpour put most of the Nawab's guns out of action, but the British gunners kept their powder dry. As the Indian artillery (manned by Frenchmen) retreated, Clive advanced to man the embankment surrounding the large village pond to the front of his position. An Indian infantry attack was repelled by Clive's artillery and infantry fire and, as the Indians retreated, Clive's men advanced rapidly, storming the Indian entrenchment. The Nawab had already fled. He was captured and murdered some time after by Mir Jafar, a key general who had reached an agreement with Clive before the battle. Mir Jaffir was installed as the new Nawab. **ABOVE**

The capture of Québec from France in 1759, which was seen at the time as a great triumph, led to the production and sale in Britain of more maps of North America. Moreover, *A universal geographical dictionary; or, grand gazetteer* (1759) was, as its title page proclaimed, 'illustrated by a general map of the world, particular ones of the different quarters, and of the seat of war in Germany'.

Readers of the *London Magazine* who wished to follow the course of the war of 1760–61 in North America between the Cherokees and British and colonial forces could turn to *A new map of the Cherokee Nation … engraved from an Indian draught* by Thomas Kitchin, in early 1760.

There could, however, be difficulties in getting hold of maps. In September 1792, at a time of rising tension as France moved in a radical direction, James Bland Burges, the Under-Secretary in the Foreign Office, noted:

I have not forgot your commission respecting the new map of France, but my endeavours to procure one for you have been unsuccessful, as I am assured by Faden, who is omniscient in all matters relating to geography, that there is not one to be met with in London. He has however promised to secure for me the first which can be found; though the uncertainty and difficulty attending the bringing over anything (except emigrants) from France must make the time of your getting what you want extremely doubtful.

Interest in cartography, of course, was not restricted to the public, but instead overlapped with military concerns. In October 1760, Lieutenant-Colonel James Montresor (1702–76) wrote to General Jeffrey Amherst, the British commander-in-chief in North America, a master of planned, methodical campaigning:

I think it my duty to acquaint your Excellency that I have got in great forwardness a general map of that part of North America which has been the seat of war wherein is distinguished the roads that have been made by the troops, the navigation of its rivers, its carrying places, the new forts and posts constructed, the several hospitals, barracks and buildings for the soldiers, the marches of the army, the places where have been engagements, attacks, sieges and camps, interspersed with useful remarks. The most part laid down by actual surveys with geographical and military observations made in that country from the year 1754 to 1760. As this map will show at one view what has been done in that country, I hope that it will be very acceptable, as well to the ministry as to the military, as your Excellency's march from Albany to Montreal, and Brigadier-General Murray's from Quebec is only wanting to complete it.

Montresor sought information on the details of these successful advances earlier in 1760. Amherst himself collected maps. Montresor's career showed the significance of British imperial conflict and expansion for mapping. His father, James Montresor, a French Protestant, had served in the British army, eventually as lieutenant-governor of Fort William in Scotland.

James Montresor himself served in the army before becoming a practitioner-engineer, gaining a reputation as a draughtsman. Having served at Gibraltar and Minorca, he became chief engineer at Gibraltar in 1747. From 1747 to 1752, Montresor produced 26 plans of various parts of Gibraltar's defence works, with sections of the fortress and barracks of Gibraltar and of the Spanish lines and forts. In 1753, he produced a plan of the fortifications.

Montresor became chief engineer of the Braddock expedition in 1755. After its disastrous defeat, Montresor prepared plans and projects, surveying and

BATTLE OF PRAGUE. BY J. H. FREYTAG, 1757 On 6 May 1757 Frederick the Great planned to roll up the Austrian position to the east of Prague from its right, but he lost control of the flow of the battle as the Austrians were able to move much of their army to cover this flank. The Prussians were therefore forced to make a frontal attack. Advancing with shouldered muskets, the Prussian infantry was repulsed with heavy losses from Austrian cannon and musket fire, and with Field-Marshal Schwerin being killed. However, the Prussians were then able to advance in the gap between the Austrian main army, still facing north, and the units that had defeated Schwerin. The exposed wing of the latter was attacked, while, at the same time, the Prussian cavalry defeated its Austrian opponents thanks to a flank attack. The principal Austrian position was indeed rolled up from its right flank, and when the Austrians rallied, they had to withdraw due to a Prussian threat to both flanks. Nevertheless, although the Austrians lost about 14,000 men, including 5,000 prisoners, the Prussians had 14,287 dead and wounded. Frederick was unable to consolidate the battle by capturing Prague. **RIGHT**

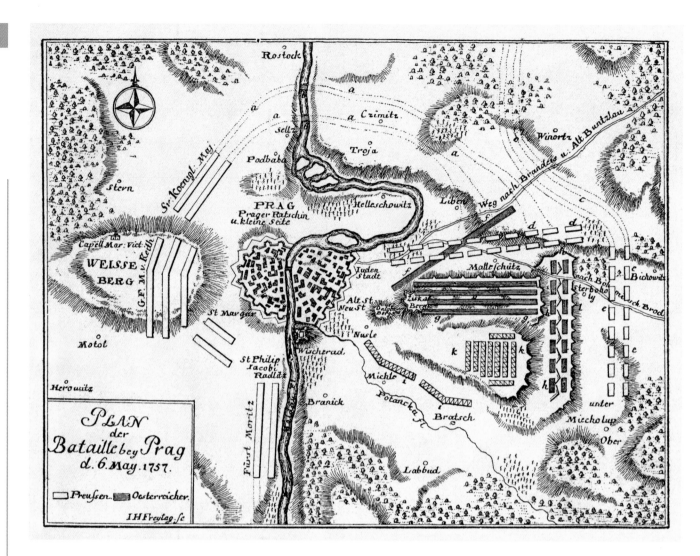

mapping part of Lake Champlain, a key area of operations where the British finally succeeded after initial failure. Becoming director of the Royal Engineers in 1758, Montresor played a leading role in operations until 1760.

Montresor's eldest son, John, a lieutenant, was wounded on Braddock's expedition. In 1758, he became a practitioner-engineer, before constructing a chain of redoubts near Niagara in 1764 and producing linked plans. John Montresor was appointed chief engineer in America in 1775 and constructed the defence lines near Philadelphia in 1777–78 after it was captured by General William Howe following the British victory at Brandywine. From this stage of his career, Montresor produced plans of Boston, New York and the battle of Bunker Hill in 1775.

British popular interest in North America after the Seven Years' War (1756–63) had continued to encourage the publication of maps, and strengthened with the War of American Independence, which broke out in 1775. Thomas Jefferys published a number of atlases, including *A general topography of North America and the West Indies* (1768). Using Jefferys' maps, Robert Sayers and John Bennett published *The American atlas, or a geographical description of the whole continent of America* (1775), with subsequent editions dated 1776 and 1778. The first major battle of the American Revolution, Bunker Hill in 1775, was rapidly followed in London by the publication of maps of the battle, the earliest appearing four days after the report of the engagement reached London.

The struggle for Boston was also followed on the

Continent, with Samuel Holland and George Callendar's *Chart of the harbour of Boston*, published by J.F.W. Des Barres, being used by Jean, Chevalier de Beaurain, to produce a map. The latter map was soon re-engraved and printed in Leipzig. More generally, the French relied in part on the availability of British printed maps sold through the map trade.

In 1776, detailed maps of New England enabled British readers, anxious about the civil war in the empire, to follow the course of the conflict in its initial theatre, while, in 1777, William Faden published a large-scale map of New Jersey, and, in the following year, a plan displaying the recent campaign on the Delaware River, which had been waged by the British in order to ensure Philadelphia's access to the sea. Faden did the same for other campaigns and engagements, such as the unsuccessful British attack on Fort Sullivan, South Carolina, in 1776 and George Washington's operations at Trenton and Princeton in New Jersey in the winter of 1776–77, operations that delivered a crucial check to the British forces after their earlier success in exploiting victory at New York in order to overrun New Jersey.

Several plans depicting the decisive clash of the war, the successful American and French siege of the British position at Yorktown in 1781, appeared shortly after the event, notably including Sebastian Bauman's masterly work, printed at Philadelphia in 1782. An interesting contrast in perspective is evident when comparing two important French maps of the siege, *Carte de la partie de la Virginie ou l'armée combinée de France et des Etats-Unis de l'Amerique a fait prisonniere l'armée anglaise* (c. 1782), which put much emphasis on the battle between the British and French fleets near the mouth of Chesapeake Bay, whereas Georges-Louis Le Rouge's map closely focused on the action immediately surrounding the besieged town, an

approach that led to a stress on the American role.

Maps, however, did not capture the key element of the political affiliations of the American public nor the difficulties facing the two sides: the British failure to create an effective pacification policy and the deficiencies of the American war machine.

There is also the need today to understand maps in terms of the communications of the period: issues of strategy were greatly exacerbated by poor communications. Before the telegraph produced the nineteenth-century communications revolution, instructions could go no faster than the swiftest horse or the speediest boat. This was an aspect of the situation in which Britain's need to confront a number of challenges around the world placed considerable burdens on its ability to control and allocate resources and to make accurate threat assessments.

KNOWLEDGE AND NEWS

The war, in turn, led to new maps that provided a prospectus for future conflicts. For example, *A new and correct map of the United States of North America* 'agreeable to the Peace of 1783' was published in New Haven in 1784. Produced by Abel Buell, it was advertised by him as 'the first ever published, engraved and finished by one man, and an American' and as designed for the 'patriotic gentleman'. The map, which presented a new land reaching to the Mississippi was a prospectus for war with the Native Americans.

The relationship between war and knowledge was repeatedly seen with cartography. Reviewing Lewis Evans' *Analysis of a general map of the middle British colonies in America* in the *Literary Magazine* of 15 October 1756, Samuel (Dr) Johnson wrote that 'the last war between the Russians and the Turks [1736–39] made geographers acquainted with the situation and extent of many countries little known before', a

BATTLE OF KUNERSDORF, 1759 This major defeat for Frederick the Great at the hands of Russia and Austria on 12 August 1759 was fought to the north-east of Frankfurt an der Oder on a site that today lies in Poland. Suffering from poor intelligence, Frederick was unable to direct the dynamic of the battle. His line of attack was poorly planned and exposed his advancing infantry to heavy artillery fire. The Prussian infantry could not sustain the heavy casualties, which totalled 19,000 men, about 40 per cent of his army. A total of 172 cannon were lost. The 48,000 Prussians were opposed by 24,000 Austrians and 40,000 Russians. Russia and Austria, however, failed to follow up the victory by concerted action. RIGHT

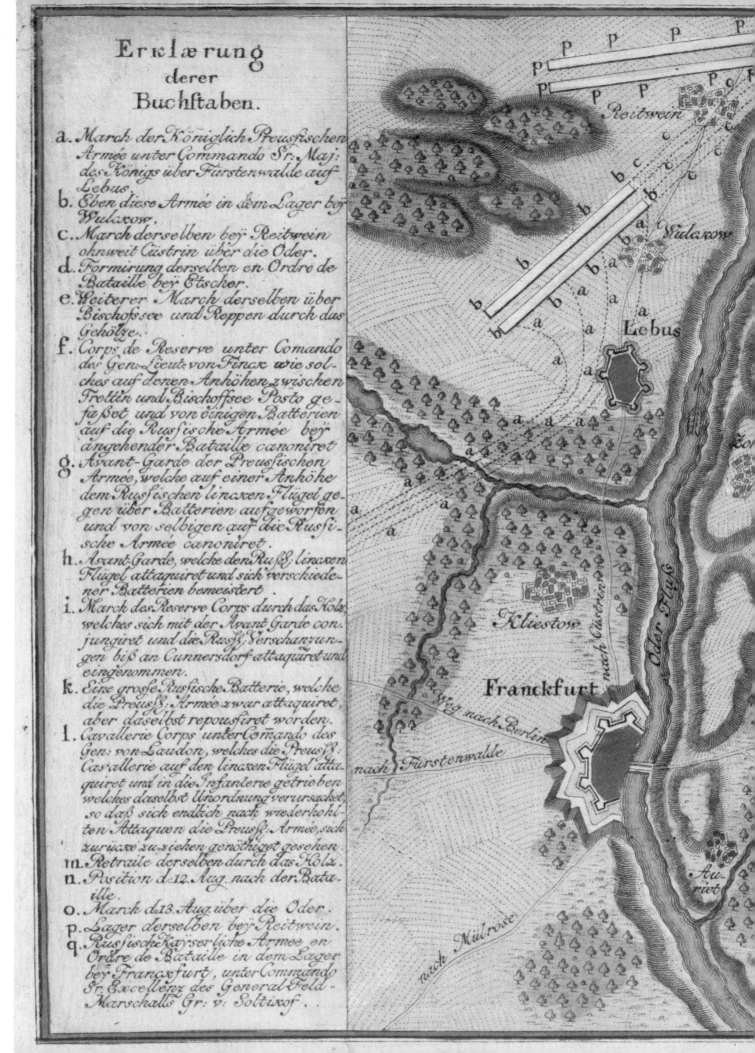

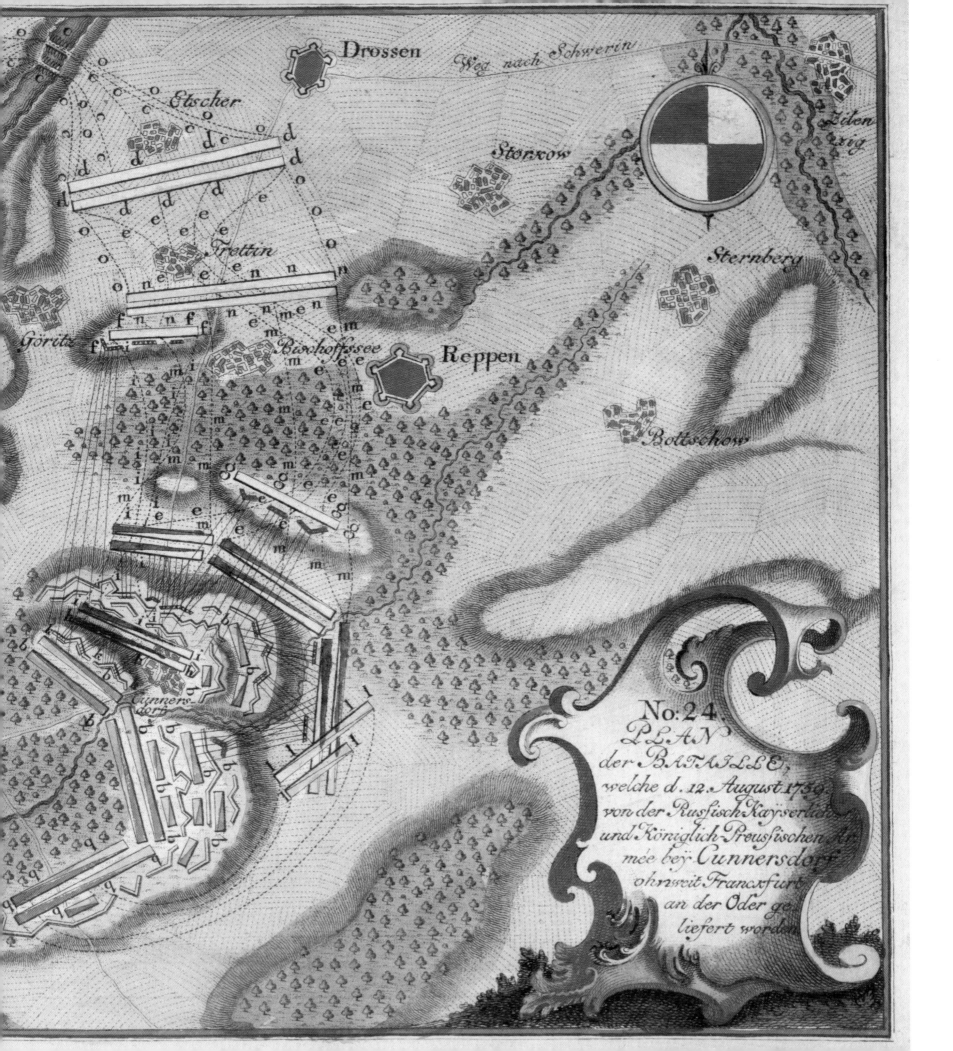

Drossen

Weg nach Schwerin

Etscher

Storcow

Idenzig

Trettin

Sternberg

Göritz

Bischoffssee

Reppen

Bottschow

Cunners-
dorf

No: 24.
PLAN
der BATAILLE,
welche d. 12. August 1759.
von der Russisch-Kayserlich,
und Königlich-Preussischen Armée bey Cunnersdorf
ohnweit Francxfurt
an der Oder ge-
liefert worden

BATTLE OF ROSSBACH, 1757 On 5 November 1757, Frederick II, known as the Great, of Prussia, with 22,000 men, attacked the 30,200-strong French and the 10,900-strong army of the Empire, which had planned to turn the Prussian left flank. Responding rapidly, Frederick attacked his slow-moving opponents on the march, screening his move behind a hill. After hard fighting, Major-General Seydlitz defeated the opposing cavalry of the French advance guard. This demoralised the infantry. The advancing columns of French infantry were brought low by salvoes of Prussian musket fire in a brief engagement, while a second charge from Seydlitz completed their collapse. The French, whose forces lacked cohesion, discipline and experience, fled in confusion. The same was true of the army of the Empire. The allies lost more than 10,000 men, mostly prisoners, the Prussians fewer than 550. Frederick's ability to grasp and retain the initiative, and the disciplined nature of his forces, were decisive. This victory, followed by that over the Austrians at Leuthen on 5 December, reversed a run of setbacks for Prussia. It made Frederick famous as a general and caused a major crisis for French reputation and self-confidence. RIGHT

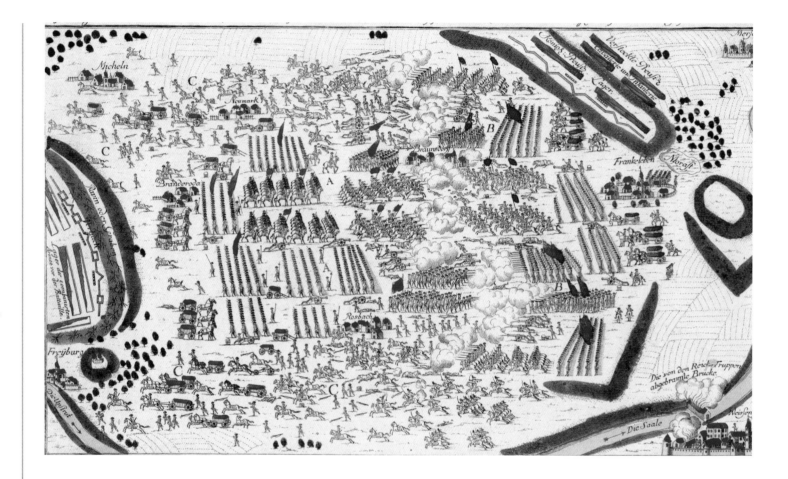

reference to the lands on the northern shore of the Black Sea: the Khanate of the Crimea, a Turkish vassal state, attracted few visitors and there were no reliable maps. However, conflict encouraged both supply and demand: military mapping and the commercial production of maps. Thus, Louis-Félix Guynement de Kéralio published *Histoire de la guerre entre la Russie et la Turquie, particulièrement de la campagne de 1769*, a book relating to the 1768–74 war, and *Histoire de la guerre des Russes et des impériaux contre les Turcs en 1736, 1737, 1738, 1739*, works that included maps. The *Journal politique de Bruxelles* of 2 February 1788 advertised a map of the northern and north-western littoral of the

Black Sea that would help those interested in the Russo-Turkish war (begun the previous year) follow its course. The net effect was a spread in knowledge about a part of the world that had been very obscure in all respects to Western Europeans earlier in the century.

In 1758, during the Seven Years' War, Lord George Sackville, who was in command of British forces in Germany, had written to Robert, 4th Earl of Holdernesse, a British secretary of state, who had travelled as a diplomat: 'You will see Cappenburgh in the map.'

The use of maps to follow wars, both in the

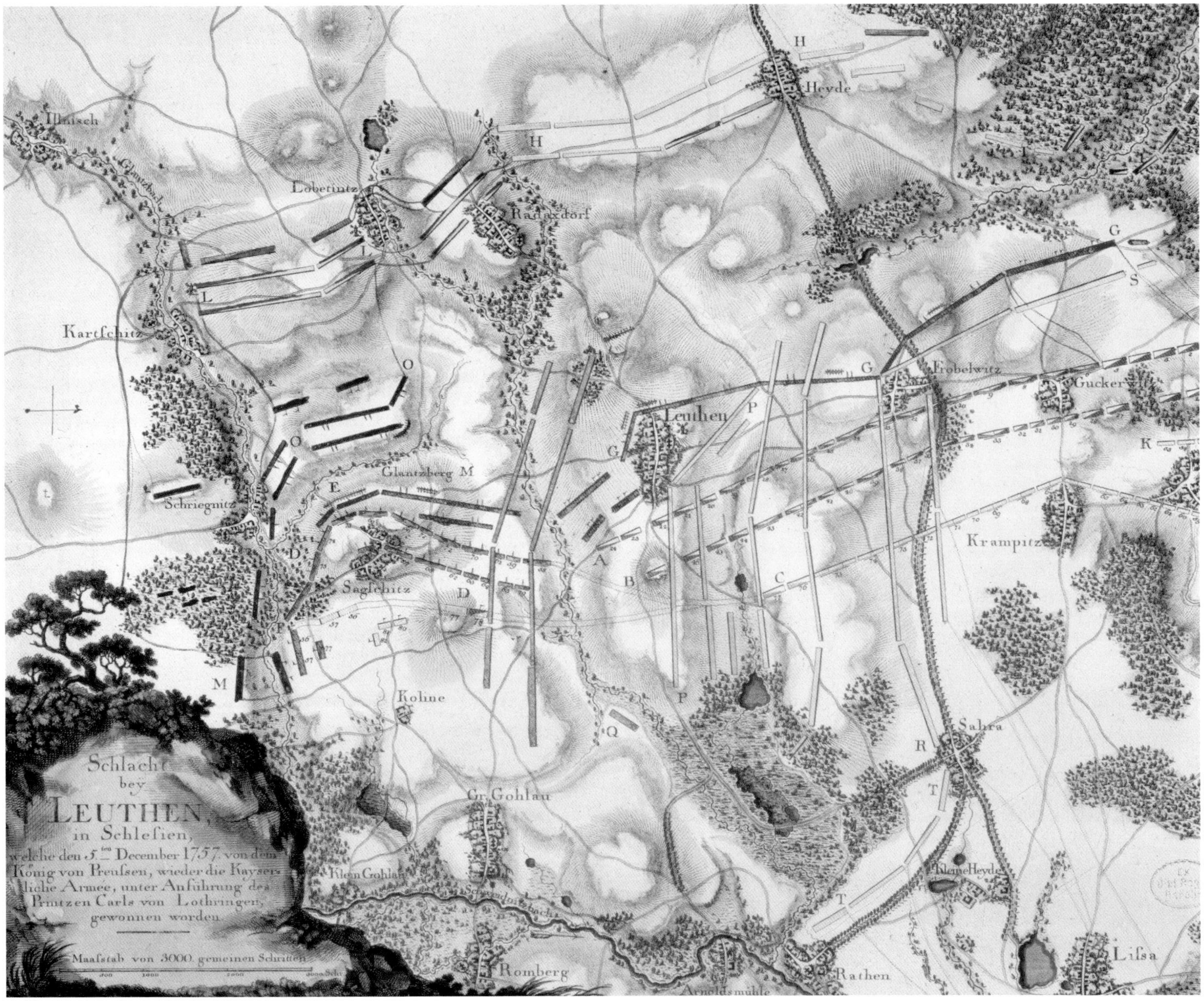
Schlacht
bey
LEUTHEN,
in Schlefien,
welche den 5.ten December 1757. von dem
König von Preufsen, wieder die Kayser-
liche Armée, unter Anführung des
Printzen Carls von Lothringen,
gewonnen worden.
Maafstab von 3000. gemeinen Schritten
Heyde
Lobetintz
Leuthen
Krampitz
Schriegnitz
Koline
Gr. Gohlau
Sahra
Romberg
Rathen
Lifsa

BATTLE OF LEUTHEN, 1757. COMMEMORATIVE MAP OF THE BATTLE PLAN MADE IN 1758 Faced by Austrian successes in Silesia, Frederick the Great advanced to Lorraine on 5 December 1757. Frederick had a first-rate army of 35,000 men, Prince Charles of Lorraine 54,000. Frederick crucially took and retained the initiative by using a fine example of the oblique order attack. Benefiting from the cover of a ridge, the Prussians turned the Austrian left flank, while a feint attack led the Austrians to send their reserves to bolster their right. The Prussians were helped greatly by mobile artillery. Charles responded to the Prussian success by wheeling his army so as to create a new south-facing front stretching through the village of Leuthen. The second phase of the battle centred on repeated Prussian attacks on this new front, especially on Leuthen, which was finally carried after bitter fighting. The Prussian infantry became exposed to the Austrian cavalry, but the latter was checked by the prompt action of the Prussian cavalry, and the battered Austrian infantry finally fled. The Prussians lost 6,380 killed and wounded, the Austrians 10,000, as well as 12,000 prisoners. After their defeat, the Austrians abandoned most of Silesia. Leuthen was a hard-fought victory by a well-honed army. It reflected Prussian firepower, Frederick's skilled exploitation of the terrain, the fighting quality of the Prussian cavalry and the ability of Prussian commanders to take the initiative. LEFT

eighteenth century and thereafter, was an aspect of a general map culture, with rising map consumption being focused on particular interests. This focus involved not only conflicts in which one's own country was a participant, but also others. Maps were part of the news explosion. Cartography served as an aspect of government applying knowledge and of the news. It was doing so in a West in which the provision of news was becoming more central, and its presentation more clearly scientific and rational, rather than providential and impressionistic.

Moreover, war was highly important both in affecting society and in leading to fundamental changes in the international system. Later, there was to be a view that *Ancien Régime* (1648–1789) warfare, that before the French Revolution in 1789, was of limited consequence, but this was not how it struck contemporaries (nor, indeed, was it the case). As far as the European world was concerned, war led first to Britain becoming the leading imperial power, and then to the independence of the United States.

China defeated the Zunghars of Xianjiang in the 1750s, overcoming the last challenge from the steppes. Chinese commanders were expected to provide maps to accompany their reports to the Emperor on wars. Maps survive from 1749 for the First Jinchuan War

*Bataille bey Hastenbeck
den 26 Jul. A. 1757.*

BATTLE OF HASTENBECK. ENGRAVING PUBLISHED IN 1758

Engraving of the French victory over a
Hanoverian and allied force under William,
Duke of Cumberland, at Hastenbeck near
Hamelin on the River Weser on 26 July 1757.
Cumberland's army fought well but was
defeated by a larger French army under
Marshal d'Estrées. That battle was followed by
the overrunning of the Electorate of Hanover
and the capitulation of Cumberland's army, by
the Convention of Klosterzeven signed on 10
September. This was an indication of the
possibility of achieving decisive victory. ABOVE

and, with greater detail, from 1800 for the White Lotus Rebellion.

In Europe, Russia became the major power in Eastern Europe, repeatedly defeating the Turks, Swedes and Poles from the 1690s and 1700s. Conscription systems, which were introduced in much of Europe during the century, greatly affected society.

MEANS OF CONFLICT

As far as the means of conflict were concerned, the emphasis in the Western world was on close-packed formations of infantry exchanging volley fire at close range. Cavalry played a major role on the flanks but was less significant than infantry in most battles. Nevertheless, cavalry breakthroughs played a key role in some battles, such as the major Anglo-Austrian defeat of the French at Blenheim in 1704. There is no systematic scholarly treatment of cavalry in this period. The attention contemporaries devoted to cavalry in part reflected the social prestige of cavalry service and the authority of its past importance. That can be gauged from paintings including the illustrations on maps but not from their diagrammatic sections. Cavalry was even less significant in sieges.

The latter remained important although their results tended to be dependent on battles. Thus, Napoleon's success in besieging Mantua, the major Austrian fortress in northern Italy, in 1796–97, was secured by his ability to defeat four Austrian relief attempts. As a result of the importance of relief attempts, maps of sieges tended to include the surrounding countryside. The terrain features that might help relief armies to attack besiegers successfully, as when the Turks were routed at Vienna in 1683 by forces successfully advancing through the Vienna woods, were highly significant.

The most important battle in Italy in the 1700s was that at Turin in 1706 when a French siege army was defeated by a relief army of Austrian and Savoy-Piedmontese forces under Prince Eugene and Victor Amadeus II. The significance of this battle ensured that it was extensively memorialised, including in maps. Thus, maps for display remained highly important, and this was the case across Europe.

The idea of natural frontiers became more significant in the late eighteenth century, supplementing or superseding earlier jurisdictional ideas of territorial identity and rights. Thus, in France, there was a commitment to the idea of mountain and river frontiers, notably the Alps, Pyrenees and Rhine. Such ideas helped ensure that the psychological idea of countries became more fixed in terms of a defined physical shape. The latter was clarified as the accuracy of maps improved and, therefore, as images became more standardised. Toward the end of the century, there were also intimations of new moves in warfare, moves that might have entailed novel needs of, and means for, mapping. In particular, manned balloon flight began in 1783 and, in the 1790s, the French used reconnaissance balloons, notably at the Battle of Fleurus south of Brussels in 1794. However, this innovation was not responsible for their success and was abandoned after Napoleon seized power in 1799. In large part, this was due to the time it took to inflate the balloons and the impossibility of controlling them. As with submarines and semaphores, it was the potential for the future that was notable, and not the situation in the 1790s. Indeed, as far as technology is concerned, changes in 1815–65, including steam, iron and shell guns, were more influential than anything from the previous fifty years or indeed the previous three centuries.

BATTLE OF ZORNDORF, FROM *UNIVERSAL*

REGISTER **MAGAZINE, 1759**   The bloody battle
at Zorndorf on 25 August 1758 was
indecisive but blocked the slow-moving
Russian invasion of Brandenburg.
Frederick had covered 250 kilometres in
twelve days in order to stop the Russians.
A protracted battle lasted about ten
hours. The Prussians lost a third of their
36,000 men and the Russians 18,000 of
their 43,500. Although helped by better
artillery and useful cavalry charges,
Frederick the Great found that he could
not control the disorganised (in the sense
of decentralised) battle nor impose a
result on the Russians, whose infantry
(like that of the Prussians) displayed great
fighting power. The battle underlined the
strain of the war on the Prussian army.
This map was published in the *Universal
Register*, a London monthly magazine, in
February 1759.   **RIGHT**

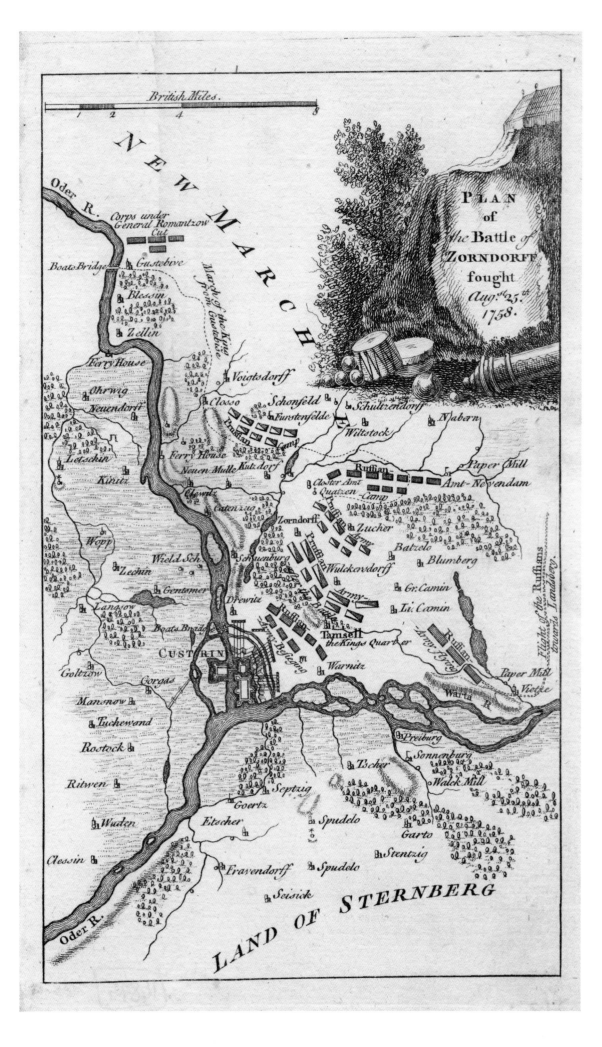

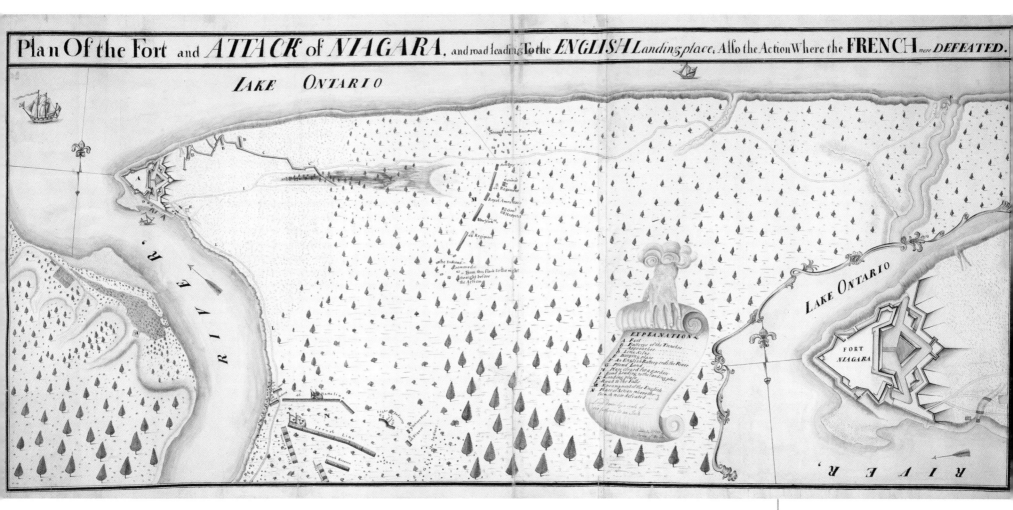

Plan Of the Fort and ATTACK of NIAGARA, and road leading To the ENGLISH Landing place, Also the Action Where the FRENCH were DEFEATED.

PLAN OF THE ATTACK ON NIAGARA, BY A BRITISH QUARTERMASTER, 1759 The French had built a fort at Niagara in 1720 to further their expansion west from the St Lawrence Valley, the heart of New France. In 1759, the British advanced to Lake Ontario from Albany on the River Hudson. Short of Native American support, the French abandoned some of their forts, but they contested the Niagara positions. However, an advancing French force was defeated by defensive fire at nearby La Belle-Famille on 24 July by a force of British regulars, American provincials and Native Americans under Colonel William Massey. The well-fortified Fort Niagara surrendered to the British two days later. As at Québec in September 1759, although on a far smaller scale, the defeat of the local French field force had been followed by the surrender of their principal position. Hope of relief had been lost. **ABOVE**

BATTLE OUTSIDE QUÉBEC, 1759 When the British forces scaled the cliffs near Québec on 13 September 1759, the French army, instead of waiting on the defensive, chose to attack. The British under General James Wolfe waited until the French were about 100 feet away, then opened regular volley fire. This led the French to retreat, their commander, the Marquis of Montcalm, mortally wounded like Wolfe. A British participant recorded: 'About 9 o'clock the French army had drawn up under the walls of the town, and advanced towards us briskly and in good order. We stood to receive them; they began their fire at a distance, we reserved ours, and as they came nearer fired on them by divisions; this did execution and seemed to check them a little, however they still advanced pretty quick, we increased our fire without altering our position, and, when they were within less than a hundred yards, gave them a full fire, fixed our bayonets, and, under cover of the smoke, the whole line charged.' Québec surrendered five days later. RIGHT

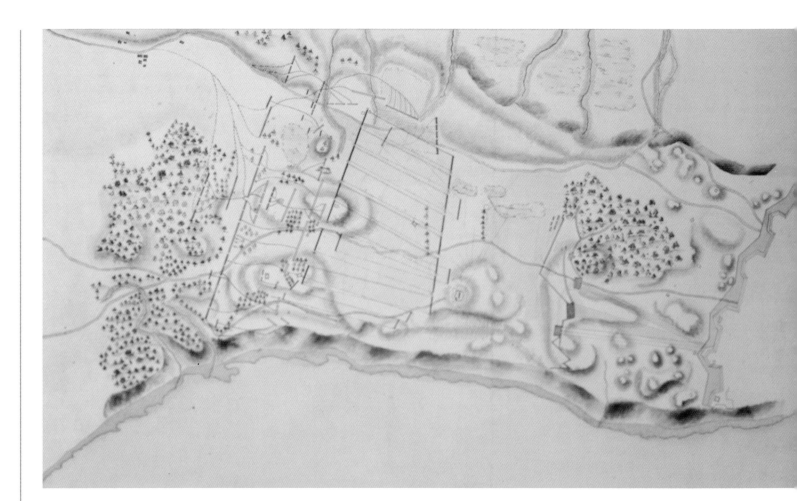

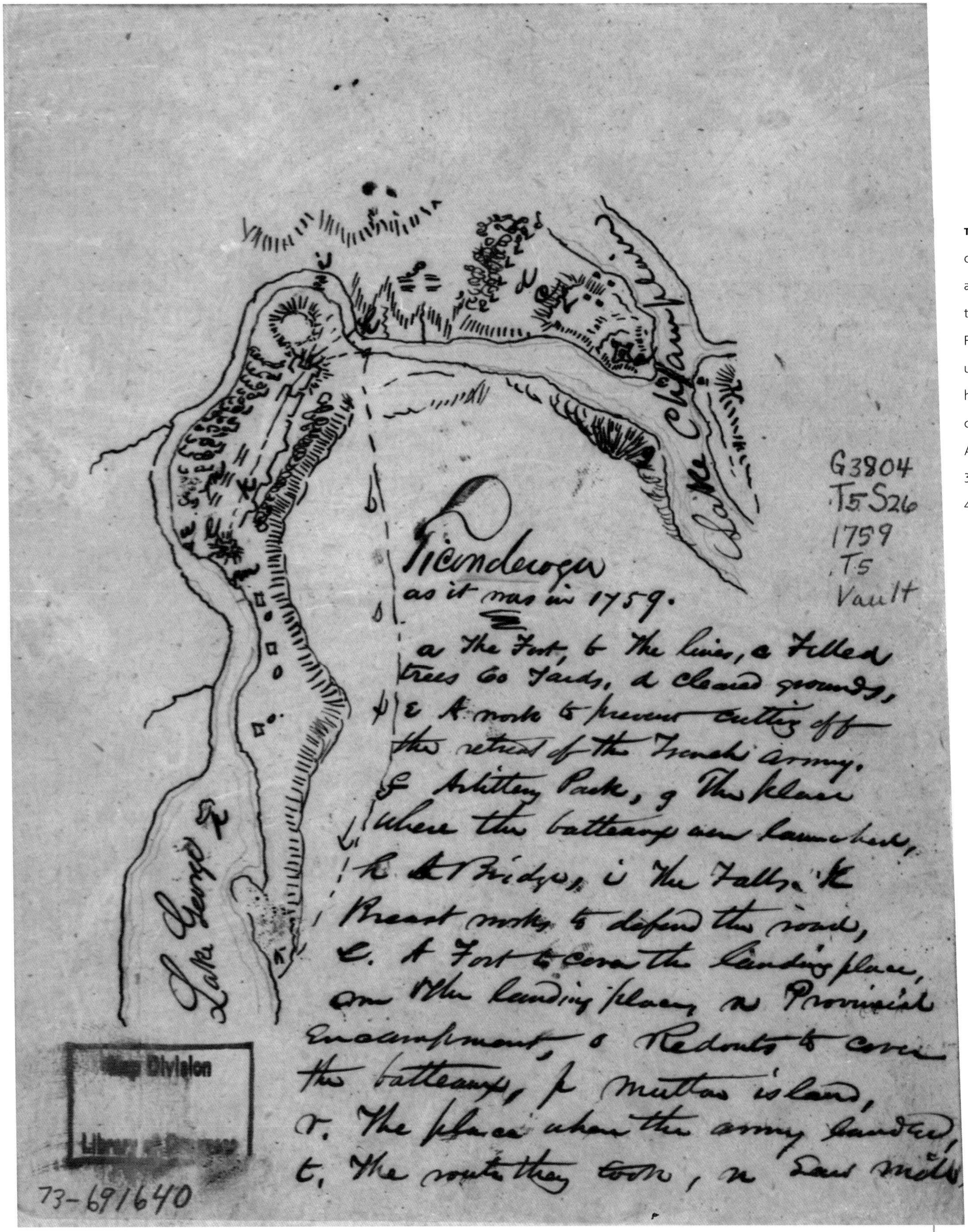

TICONDEROGA, AS IT WAS IN 1759 A hand-drawn map that outlined the situation and terrain. Advancing on Canada from the south, the British had attacked the French position the previous year, with an unsuccessful frontal assault leading to heavy casualties; 1,900 killed or wounded out of a force of 6,400 regulars and 9,000 American provincials. In contrast, the 3,500-strong French force suffered only 400 casualties. **LEFT**

PLAN FOR ATTACK ON TICONDEROGA. BY WILLIAM BRASIER, 1759 This map shows the details of the planning for an attack and the determination to link it to an operational grasp of the environs, including securing communications and carrying boats overland. In the event, the simultaneous British attack on Québec led the French to respond to the British advance by abandoning the Ticonderoga position on the night of 26–27 July. It was then occupied by the British. Unprepared, Ticonderoga fell to an American force advancing from the Hudson Valley in May 1775, being easily regained by the British in July 1777 when superior strength was brought up from Canada. The Americans had failed to fortify Mount Defiance, which commanded their positions, because they believed that it was impossible to get artillery up its slopes. Appreciating its value, the British moved artillery onto the mountain, leading the Americans to abandon the position. RIGHT

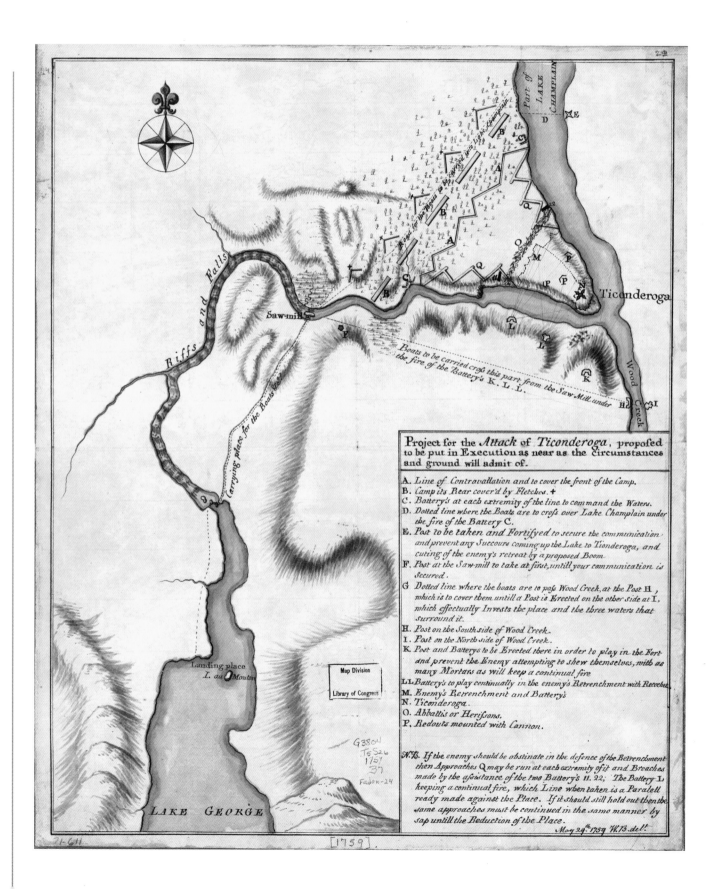

Project for the *Attack* of *Ticonderoga*, proposed to be put in Execution as near as the Circumstances and ground will admit of.

A. Line of Contravallation and to cover the front of the Camp.
B. Camp its Rear cover'd by Fletches. +
C. Battery's at each extremity of the line to command the Waters.
D. Dotted line where the Boats are to cross over Lake Champlain under the fire of the Battery C.
E. Post to be taken and Fortifyed to secure the communication and prevent any Succours coming up the Lake to Tionderoga, and cuting of the enemy's retreat by a proposed Boom
F. Post at the Saw-mill to take at first, untill your communication is secured.
G. Dotted line where the boats are to pass Wood Creek, at the Post H, which is to cover them untill a Post is Erected on the other side at I, which effectually Invests the place and the three waters that surround it.
H. Post on the South side of Wood Creek.
I. Post on the North side of Wood Creek.
K. Post and Batterys to be Erected there in order to play in the Fort and prevent the Enemy attempting to shew themselves, with as many Mortars as will keep a continual fire
L.L. Battery's to play continually in the enemy's Retrenchment with Retrchet
M. Enemy's Retrenchment and Battery's
N. Ticonderoga.
O. Abbattis or Herissons.
P. Redouts mounted with Cannon.

N.B. If the enemy should be obstinate in the defence of the Retrenchment then Approaches Q may be run at each extremity of it, and Breaches made by the assistance of the two Battery's 11. 22; The Battery L keeping a continual fire, which Line when taken is a Paralell ready made against the Place. If it should still hold out then the same approaches must be continued in the same manner by sap untill the Reduction of the Place.

May 29th 1759 W.B. del.

SIEGE OF HAVANA, 1762 A British force of 12,000 troops under George, 3rd Earl of Albemarle, covered by twenty-two ships of the line, landed to the east of Havana on 7 June 1762. Operations were concentrated against the still-impressive Fort Moro which commanded the channel from the sea to the harbour and was protected by a very deep landward ditch. On 1 July, the British batteries opened fire, supported by three warships, but damage from Spanish fire forced the latter to abandon the bombardment. The summer passed in siege works, which were hindered by the bare rock in front of the fortress, and in artillery duels. A third of the British force was lost to malaria and yellow fever, but the Spanish batteries were silenced by heavier British fire. On 30 July, the British exploded two mines on either side of the ditch, creating an earth ramp across it and a breach that was stormed successfully, enabling them to capture the fort. From there, British artillery could dominate the city, and it surrendered on 13 August. The fleet in the harbour, which included twelve ships of the line, also surrendered. LEFT

BOSTON

CHARLES TOWN

BATTLE OF BUNKER HILL, 1775 On the night of 16–17 June 1775, the Revolutionaries marched to Breed's Hill on the Charlestown peninsula and began to fortify the position which commanded the heights above the city of Boston; the battle is named after the more prominent hill behind Breed's Hill. The British decided on moving to the peninsula on a high tide that afternoon, followed by an attack on the American entrenchments. However, they moved ponderously, spending about two hours deploying and then advanced in a traditional formation. The British artillery failed to damage the American positions significantly. The Americans waited until the advancing British were almost upon them before shattering their first two attacks with heavy musket fire. An attempt to turn the American flank was repelled. The Americans, however, were running short of ammunition and a third British attack took the American redoubt. The exhausted British, harassed by sharpshooters, were unable to stop the Americans from retreating. British casualties were heavy: 228 dead and 826 wounded, 42 per cent of their force engaged, compared with 100 Americans killed and 271 wounded. Thereafter, the British army in Boston was singularly cautious in mounting operations. As a result, the British lost the strategic and operational initiatives; their troops did no more than control the ground they stood on.
ABOVE

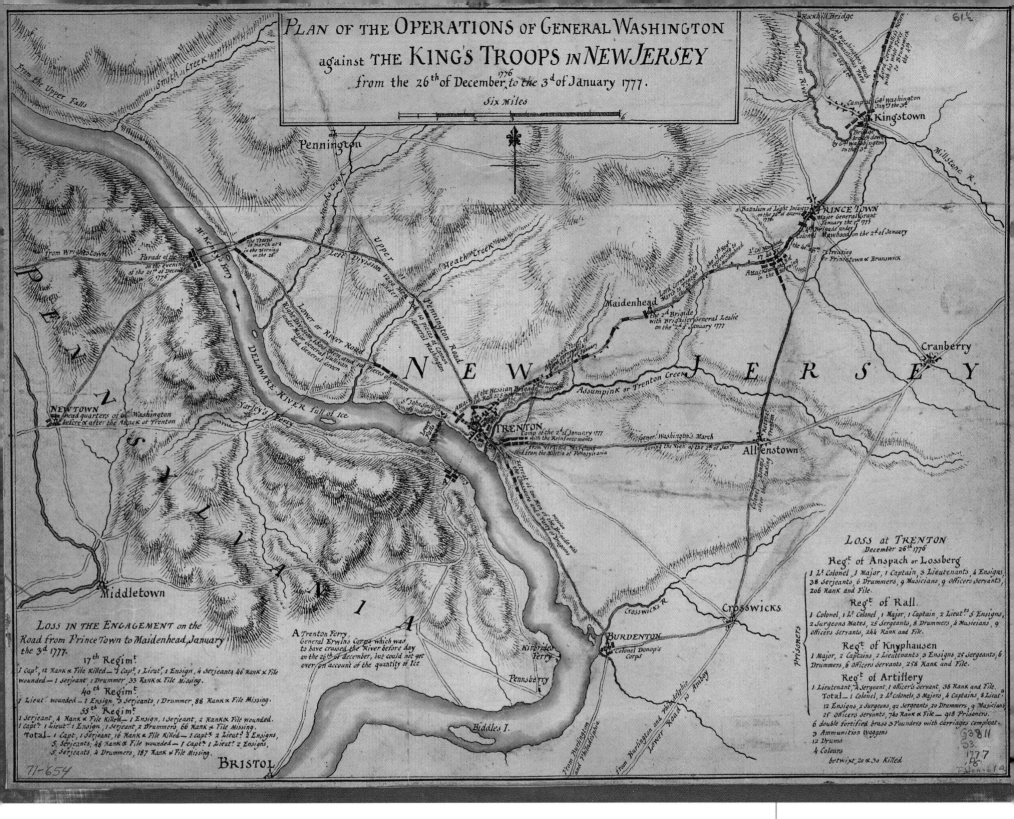

PLAN OF THE OPERATIONS OF GENERAL WASHINGTON against THE KING'S TROOPS IN NEW JERSEY from the 26th of December, 1776, to the 3d of January 1777.

Six Miles

AMERICAN OPERATIONS IN NEW JERSEY, 1777

Washington's night crossing of the icy Delaware River at Christmastime 1776 and the surprise attack on the Hessian post at Trenton proved a brilliant success. It was followed by Washington outthinking a British reinforcing force under Cornwallis. On 3 January 1777, a British force south-west of Princeton (Prince Town on the map) was beaten by the Americans. Washington, however, was unable to follow up this victory by attacking New Brunswick, because of expiring enlistments, the tiredness of his troops and British strength. He retired to the hilly country around Morristown to camp for the remainder of the winter. Washington's victories helped revitalise resistance and permit Congress to raise a new army in 1777. ABOVE

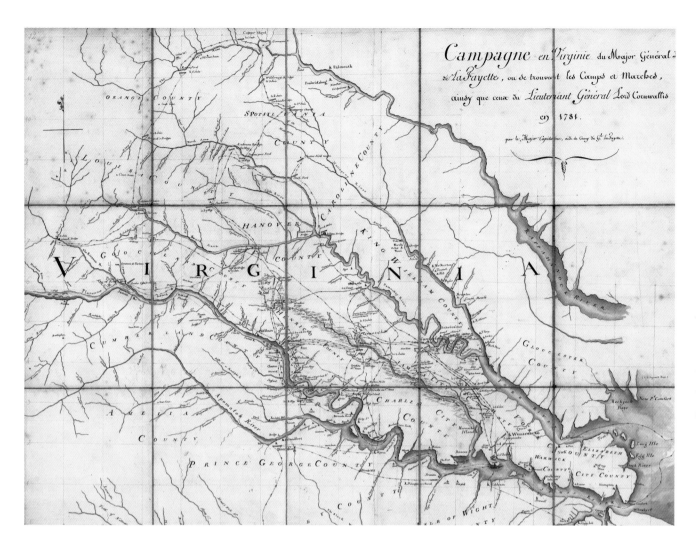

CAMPAIGN OF VIRGINIA BY THE FRENCH UNDER LAFAYETTE, BY CAPTAIN MICHEL DU CHESNOY, 1781

French military map-makers were highly impressive. The ability in this campaign to concentrate French and American forces displayed a high level of operational coordination, one that combined land and sea capabilities. The latter was particularly important for moving siege guns which were scarcely mobile for any distance. On 19 August, troops from Washington's army destined for Virginia began crossing the Hudson. On 1 September, 3,000 French troops from the West Indies landed near the entrance of the James River. Others came from New England. On the night of 28 September, American and French forces took up position round Yorktown. About 16,000 troops were moved to besiege Yorktown. The map brings out the role of river features. **LEFT**

SIEGE OF YORKTOWN, BY GEORGES-LOUIS LEROUGE, 1781 This map reflected interest in Paris in this impressive Franco-American victory of 1781 and captured the dynamic nature of operations, not least the way in which the British had fallen back while their eventual position was exposed to heavy artillery fire. The British abandoned their outer works on the night of 29 September in order to tighten their position in the face of the more numerous besiegers. The besiegers began to build the first parallel, a trench parallel to the fortifications, on the night of 6 October and the second and closer one on 11 October. On the night of 14 October, the besiegers bravely stormed the two redoubts that obstructed the path of the second parallel to the river. Two days before the surrender on 19 October, Johann Conrad Döhla, a member of the Ansbach-Bayreuth forces in the British army, recorded: 'At daybreak the enemy bombardment resumed, more terribly strong than ever before. They fired from all positions without let-up. Our command, which was in the Hornwork, could hardly tolerate the enemy bombs, howitzer, and cannonballs any longer. There was nothing to be seen but bombs and cannonballs raining down on our entire line.' **RIGHT**

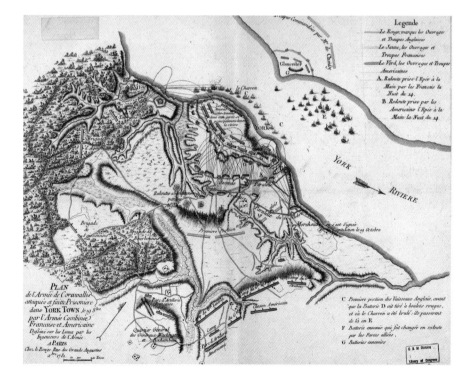

THE GORDON RIOTS, 1780 Mapping served to help visualise the issue of controlling London. Pressure from the Protestant Association under Lord George Gordon for the repeal of the 1778 Catholic Relief Act led to a challenge to order in the centre of empire. On 2 June 1780, Parliament refused to be intimidated into repeal and the demonstrators turned to attack Catholic chapels and schools, before threatening establishment targets such as the houses of prominent ministers and politicians thought to be pro-Catholic and of magistrates who sought to act against the rioters. The prisons, notably Newgate and the Clink in Southwark, were stormed in order to release imprisoned rioters, while distilleries and breweries were pillaged and the Bank of England threatened. Angered by the magistrates' refusal to act, an unflustered George III, 'convinced till the magistrates have ordered some military execution on the rioters this town will not be restored to order,' summoned the Privy Council which empowered the army to employ force without the prior permission of a magistrate. George then sent in troops to end the riots and they rapidly did so. The map shows the disposition of the troops and the general route of the patrols. ABOVE

MAP AS CARICATURE: ENGLAND WITH THE FACE OF JOHN BULL. BY JAMES GILLRAY, PUBLISHED BY H. HUMPHREY, 1793 The French threat is to be dispersed with excremental force. Satirical maps were not common in the eighteenth century, but became more so subsequently. So also did the inclusion of maps in caricatures. In 1803, Charles Williams produced his caricature *The Governor of Europe, stopped in his career.* This caricature depicts Napoleon striding across a map of conquered Europe, only to be repelled by John Bull, who defends the British Isles and has cut off Napoleon's toe with a sword. **RIGHT**

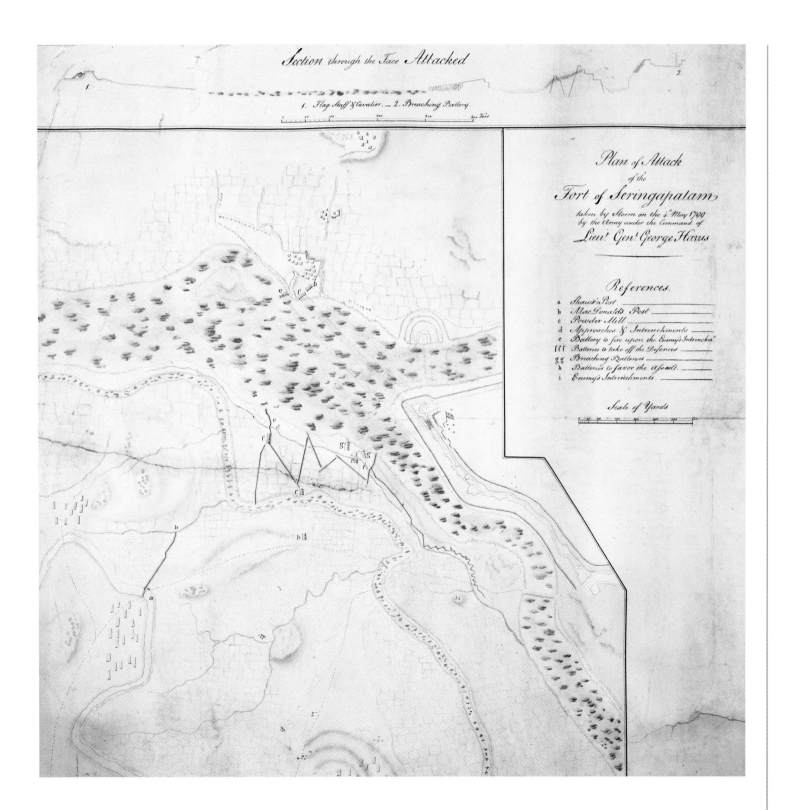

Section through the Face Attacked

1. Flag Staff & Cavalier. — 2. Breaching Battery

Plan of Attack
of the
Fort of Seringapatam
taken by Storm on the 4 May 1799
by the Army under the Command of
Lieut Genl George Harris

References.
a   Shawe's Post
b   MacDonald's Post
c   Powder Mill
d   Approaches & Intrenchments
e   Battery to fire upon the Enemy's Intrenchm'
fff Batteries to take off the Defences
gg Breaching Batteries
h   Batteries to favor the Assault.
i   Enemy's Intrenchments

Scale of Yards

ATTACK ON SERINGAPATAM. 1799   The well-defended Mysore capital, a formidable position on an island in the River Cauvery in southern India, had resisted British attack in 1791, only to surrender the following year. In 1799, in a new war, the British advanced anew, fighting off an attack by Mysore forces at the Battle of Mallavelly on 27 March, before beginning the siege on 5 April. Major-General George Harris had to succeed before the monsoon swelled the river. The British artillery on the opposite bank blew a breach in the ramparts. This was then stormed under heavy fire on 4 May. Part of the British force was held in savage fighting until the defenders were outflanked by British troops who had gained the inner rampart and then moved along. The defenders were then thrown into disorder and slaughtered with heavy losses, including the ruler, Tipu Sultan. The rest of Mysore rapidly surrendered.

LEFT

MAPPING *in an* IMPERIAL AGE

BATTLE OF MARENGO, 1800 Like many, this was a battle in which a capacity to respond to the unexpected and to fight through was crucial. On 14 June 1800, the Austrians proved a formidable rival and Napoleon's enforced retreat for much of the battle was only reversed because of a successful counter-attack mounted by French reinforcements. The fighting quality of the experienced French forces proved important, not least the ability to keep going in adverse circumstances. Thanks in part to the favourable spin he gave to his eventually successful generalship at Marengo, Napoleon's grasp of power was cemented. **RIGHT**

BIRD'S-EYE VIEW OF THE GETTYSBURG BATTLEFIELD. BY JOHN B. (JOHN BADGER) BACHELDER, (1825–1894), 1863 The map shows the positions of Union and Confederate armies during the battle and was printed c.1863. **PREVIOUS PAGES**

THE PERSISTENT WARFARE in Europe and the European world from 1792 to 1815 encouraged mapping, adding reasons for urgency in producing maps and making them available. The reasons for this mapping varied, reflecting the degree to which maps served a range of needs in war. Moreover, the wars of the period ranged widely. Russian forces operated in Holland, France, Switzerland, Finland, the Balkans and the Ionian Islands, while French forces operated from Portugal to Moscow.

ORDNANCE SURVEY

In what rapidly became a bitter struggle for survival against spreading French power, Britain was at war with France from 1793 to 1802 and 1803 to 1814 and again in 1815. One of the most lasting legacies was the Ordnance Survey in Britain. The Board of Ordnance, the government department responsible for the artillery, was given responsibility for mapping in order to help cope with a possible French invasion, which, indeed, was planned. This process predated the wars, although it was greatly encouraged by it.

The Trigonometrical, later called Ordnance, Survey – the basis of the detailed maps of Britain to this day – began in 1791, although it continued triangulation work first begun in 1784 for peaceful purposes, in order to link the observatories of London and Paris. The Corps of Royal Military Surveyors and Draughtsmen was founded in 1791 and, by 1795, they had completed a double chain of triangles from London to Land's End. A one-inch-to-the-mile map of Kent, a key invasion area, was published in 1801. Indeed, in 1804–05, French troops were deployed ready to invade as soon as naval superiority in the Channel could be obtained. In the event, the French were thwarted by the British navy. After 1801, the surveyors moved to the south-west of England,

another potential invasion area, which was covered by about 1810.

TOPOGRAPHY

There was a considerable emphasis in the mapping of the period on depicting 'strong ground': terrain that could play a role in operations. This emphasis reflected the importance of relief and slopes, not only to help or impede advances, but also for determining the sight lines of cannon. They became more important in field operations from the 1790s as the major improvements made in the French artillery from the 1760s, notably in response to failure in the Seven Years' War, affected campaigning in Europe. Napoleon, in particular, made adroit use of artillery. Direct fire was the key element for cannon, so it was necessary to understand terrain.

The mapping of topography was to be more common and more precise by the late nineteenth century, by which time the surveying and mapping of height had improved, not least with the use of contours. Nevertheless, there were useful devices prior to that. One such was the use of numbers to indicate the relative height of the ground. Important to locating artillery, this technique, known as relative command, was taught by François Jarry, a French refugee who became topographical instructor at the newly established British Royal Military College in 1799. Jarry influenced the teaching of reconnaissance to the Royal Engineers, and this helped Arthur, Duke of Wellington, as he planned in 1815 how best to respond to a French invasion of Belgium.

Relative command was used on what is termed the Waterloo map. It survives in the Royal Engineers Museum at Gillingham with pencil marks reputably made by Wellington that reveal his plans for troop dispositions. The map was marked up by Wellington the day before the Battle of Waterloo and given by

BATAILLE DE MARENGO

Iʳᵉ PLANCHE.

L'armée Autrichienne ayant débouché de ses ponts de la Bormida, et repoussé du poste de Pedro-Bona la division Gardanne qui formait l'avant-garde de l'armée Française, se déploye sur deux lignes, la droite, sous le Général Haddick, appuyée à la Bormida et la gauche, sous le Général Kaim, se prolongeant obliquement sur les bords du ruisseau.

Sa nombreuse cavalerie, sous le Général Elnitz, se porte en colonne au Nord de Castel-Ceriolo; elle est suivie des chasseurs Tyroliens et du Loup destinés à occuper ce village, soutenus par une partie des Grenadiers de la Réserve.

Ce dernier corps marche en colonne sur la grande route, sous le commandement du Général Ott.

Les deux divisions Françaises, Chambarlhac et Gardanne, sous le commandement du Général Victor, sont en position sur la droite du chemin de Marengo, occupant fictement ce village; leur droite est soutenue par la brigade de cavalerie du Général Champeaux, et leur gauche par celle du Général Kellermann.

La division Watrin, et la brigade d'infanterie du Général Mainoni, viennent à la droite de cette première ligne sous les ordres du Lieutenant Général Lannes.

Déjà l'artillerie a commencé son feu ainsi que la mousqueterie des avant-postes.

La Garde de Bonaparte est en réserve, en avant de la ferme de la Buzana.

La division Monnier, sur la droite en arrière, marche en colonne et déjà paraissent, sur la route de San-Giuliano, les éclaireurs de la division Desaix.

Ces mouvemens ont lieu de 8 à 10 heures du matin.

EXPLICATION DES SIGNES

	Positions.		
	Actuelles.	Intermédiaires.	Anciennes.
Infanterie de ligne			
Infanterie légère			
Cavalerie			
Dragons			
Chasseurs			
Husards			
Quartier-général	★	☆	☆
Infanterie			
Cavalerie			
Marche en avant			
Marche en retraite			
Lignes à l'Italienne			

Echelles

de 4 Millimètres pour 100 Mètres

de 4 Lignes pour 100 Toises

BATTLE OF TALAVERA, 1809. BY MAJOR GEORGE

HARTMANN The French under Marshal Victor attacked the Anglo-Spanish army at Talavera on 28 July 1809, concentrating their assault on the outnumbered British. Arthur Wellesley (later Duke of Wellington), the British commander, employed his infantry firepower, as at Vimeiro in 1808, to repulse the French columns. However, the pursuit of the retreating French threw the British into confusion, and fresh French units drove them back. The final French attack on the centre of the Allied line was only just held, with Wellesley committing his reserve, but held it was, although the British suffered 5,400 casualties, more than a quarter of the force. However, as a reminder of the operational context of tactical achievement, Wellesley, himself hit by logistical problems, subsequently had to retreat to Portugal in the face of fresh, larger French forces under Marshal Soult, who advanced on his lines of communication. In Parliament, Charles II, 2nd Earl Grey, for the opposition, described Talavera as a disaster, while William Windham compared it to Edward III's victory over the French at Crécy in 1346. RIGHT

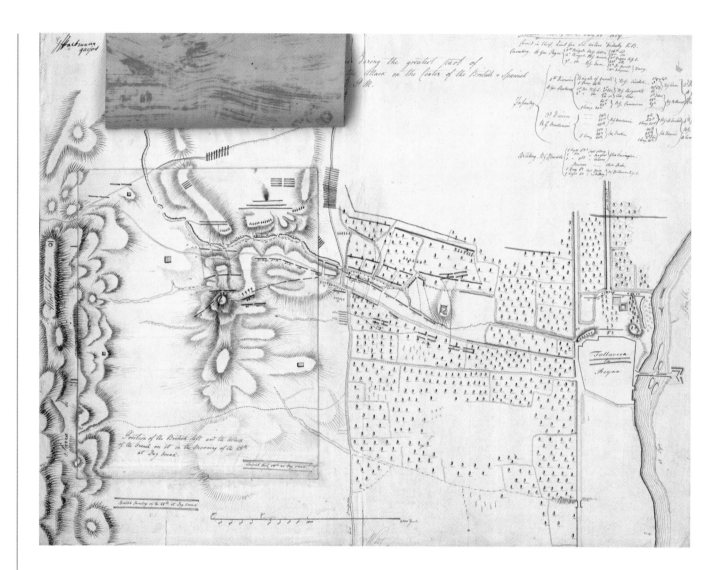

him to Colonel Sir William De Lancey, the Quartermaster-General, with instructions to deploy the troops accordingly. De Lancey was carrying the map when he was mortally wounded by a French cannonball while in Wellington's entourage near the close of the battle. Earlier, Wellington had used a mobile lithographic printing press in the Peninsular War (in Portugal and Spain) in 1808–13 in order to produce maps.

The British were not alone in developing the

mapping of height. Major-General Karl von Müffling, who became Chief of the Prussian General Staff in 1821, developed a system marking the gradients of hills on maps which became known as the Müffling method. He later introduced the optical telegraph into the Prussian army.

GOVERNMENT USE

European governments had ready access to maps. In 1800, George Canning, later a successful Foreign

Secretary, wrote to his successor as Under-Secretary in the British Foreign Office about French campaigning in Italy where Napoleon was in command of the French forces: 'What do you think of the Italian news? And what consolation does Pitt point out after looking over the map in the corner of his room by the door?' This was a reference to the Prime Minister, William Pitt the Younger. Maps were preferred to globes as they offered more detail. At the European level, they were more utilitarian than globes, which, in contrast, had a partly symbolic function alongside their value in depicting the world as a whole.

Drawing on her *ancien régime* tradition, Britain's principal opponent, France, also used maps extensively. For example, Napoleon's initially successful invasion of Egypt in 1798 led to the first accurate map of the country. The French army carried out the trigonometric survey of the Nile valley and the Mediterranean coasts of Sinai and Palestine, all of which were areas in which it operated. The resulting map, on 47 sheets, was designed to help military planning, as well as being part of a geographical enquiry that was intended to ennoble Napoleon in enlightened European opinion.

BATTLE OF WATERLOO, 1815. BY WILLIAM DUNCAN, 1836 Such maps did not capture the dynamics of the battle. Wellington's force, about 68,000 strong (31,000 of them British), was deployed on a frontage of three kilometres. The majority of his troops were deployed west of the Brussels–Genappe road. In contrast, Wellington's left, to the east of this road, was less strongly protected, both in the main frontage and, even more, in flank units. This lack of strength was due to the position of the Prussians to the east and to the Prussian promise to march to Wellington's assistance. The unevenness of Wellington's army helped encourage him to rely on a defensive deployment largely anchored on British units. Napoleon proved a poor commander, unable to make combined operations work. LEFT

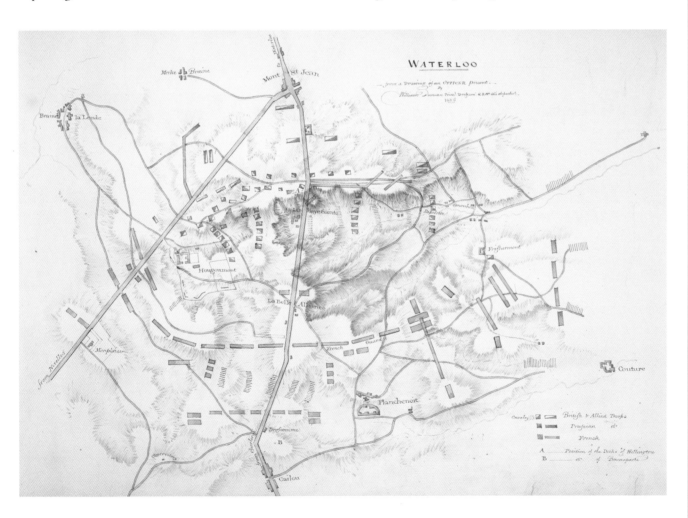

DIORAMA OF WATERLOO, BY CAPTAIN WILLIAM SIBORNE, 1815 This part of Captain William Siborne's diorama of Waterloo displays troops positioned at 19.45 on 18 June 1815. Siborne, who had not been present at Waterloo, undertook the construction of a model. He did so on the basis of thorough and lengthy research, living for eight months at the farmhouse of La Haye Sainte, a key point in the battle. Siborne also sent detailed questionnaires to British officers who had taken part, questionnaires that resulted in markedly differing answers. In 1833, the Whig government refused to allot funds for the work and Siborne did not recoup the £3,000 the model had cost. Siborne tried to portray the action at what he saw as the moment of 'crisis' in the battle, the repulse of the French Imperial Guard. Wellington regarded the model as flawed, both conceptually and methodologically. The Prussians were shown too far forward. **ABOVE**

In addition, the French occupation led to the production of a map of Cairo in which the French had a garrison. That French rule of the city included a heavy bombardment in response to a rebellion there indicating the range of activity that this information was designed to further. Artillery again emerged as a key element in control.

More generally, Napoleon, who, having seized power in a coup, was in charge of the French government from late 1799 to 1814 and again in 1815, regarded maps as a key operational tool. In 1809, he created an Imperial Corps of Surveyors with a staff of ninety. In his frequent criticisms of his subordinates, Napoleon often commented that, if they simply looked at a map, they would see the error of their ways. Describing his headquarters in 1813, when Napoleon was fighting hard, and ultimately unsuccessfully, against the Prussians, Russians and, finally, also the Austrians, to hold on to his dominant position in Germany, Baron Odeleben wrote:

In the middle … was placed a large table, on which was spread the best map that could be obtained of the seat of the war…. This was placed conformably with the points of the compass … pins with various coloured heads were thrust into it to point out the situation of the different corps d'armée of the French or those of the enemy. This was the business of the director of bureau topographique, … who possessed a perfect knowledge of the different positions … Napoleon … attached more importance to this [map] than any want of his life. During the night … was surrounded by thirty candles… When the Emperor mounted his horse, … the grande equerry carried [a copy] … attached to his breast button … to have it in readiness whenever [Napoleon] … exclaimed 'la carte!'.

The description offers an instructive guide to the significance attached to corps. On the morning of Waterloo on 18 June 1815, Napoleon briefed his commanders with maps spread over the table before him. More generally, maps were both highly useful and a key prop of power, one of the visual displays of skill in command. It was as if the ability to read the battlefield conveyed the mystery of success.

AFTER NAPOLEON

Mapping became even more significant in the post-Napoleonic world. In part, this reflected the development of cartographic and surveying experience during the Napoleonic Wars and in part continuing earlier trends in mapping within Europe. Thus Müffling noted in his memoirs:

Napoleon had commissioned Colonel Tranchon to survey the four 'Département réunies'. He established a network of principal triangles along the Rhine and Meuse, and there connected them with the great Crayenhof triangles, which stood in connection with the French measurement for ascertaining the value of a degree terminating at Dunkirk, and these again with the two bases measured in England. Tranchon had been interrupted in his surveys by the war. Already, at the First Peace of Paris [1814], we had stipulated that this map should be delivered to us; however, this was not done till the Second Peace in 1815. I had retained it in Paris, and sketched out a plan for completing it, which was approved by the King [Frederick William III] and fully accomplished by 1818. I was empowered to employ a number of hopeful young officers in surveying during the summer, and in drawing in classes during the winter, and to combine these with military instruction in staff knowledge.

In 1827, Vicente López painted a portrait of Pedro, 13th Duke of Infantado (1768–1841), the Captain

SEVASTOPOL DURING THE CRIMEAN WAR, 1853–56

Both a map and a panoramic view, this shows the north side of the Russian base from Eupatoria, with the new Russian forts and earthworks. Besieging Sevastopol, the main city in Crimea, in 1854–55 led to the Allies losing mobility, while the siege was not fully effective as road links to the north of the port remained open. This was a consequence of the lack of sufficient Allied troops to mount a comprehensive blockade. Sevastopol, a key naval base, was well defended by strongly entrenched Russian forces and the attacking Allied armies lacked adequate experience in siege craft and had to face a type of trench warfare. In the end, Sevastopol fell after a successful French surprise attack on the Malakoff Redoubt, a key position in the defences. **RIGHT**

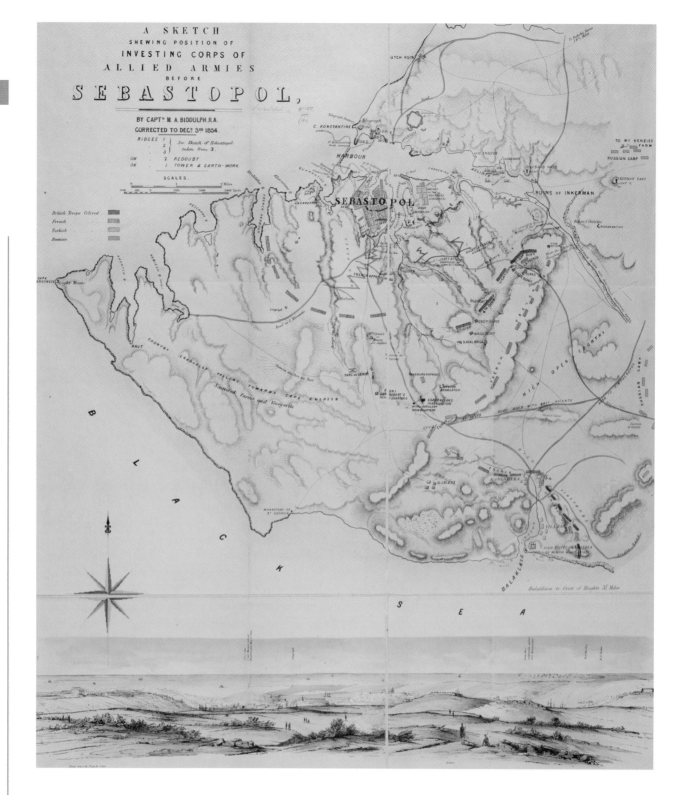

General and former Prime Minister, as pointing at a map of Spain in order to help in planning his campaigns.

Much of the warfare of the period was linked to the expansion of the West, in Africa, Asia, Australasia, Oceania and the Americas, an expansion in which military activity, including war, repeatedly played a key part. This expansion ensured a need for new maps, and maps offering considerable detail. Mapping was significant for conquest, for maintaining power, and for giving a shape to the territories over which sovereignty was asserted and enforced.

INSTITUTIONAL FRAMEWORK

The military took a major role in mapping, although other agents of the imperial state were also expected to help. In India, British army officers played the leading part. William Lambton, for example, began the

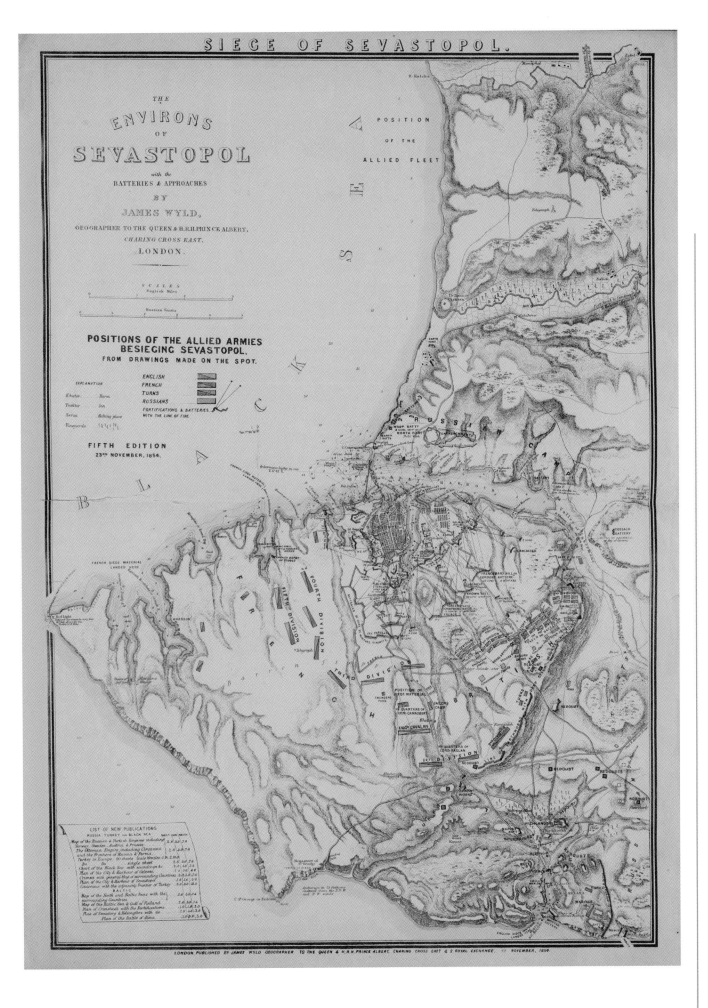

ENVIRONS OF SEVASTOPOL, BY JAMES WYLD, 1854

James Wyld was one of the great entrepreneurial map-makers of the period. The map makes clear the dependence of the Allied forces on their supply route to the harbour at Balaclava to the south. Their maritime links were vital to their successful deployment. This dependence was the cause of the battles of Balaclava and Inkerman in which the Russians were blocked. At Inkerman, on 5 November 1854, advancing Russian columns seeking to close to bayonet point took heavy losses from the Enfield rifles of the British, which showed that the formations and tactics of Napoleonic warfare were not longer possible except at the cost of heavy casualties. This lesson was to be underlined during the American Civil War (1861–65). However, Allied strategy occurred almost by accident; there was an absence of purposeful planning. **LEFT**

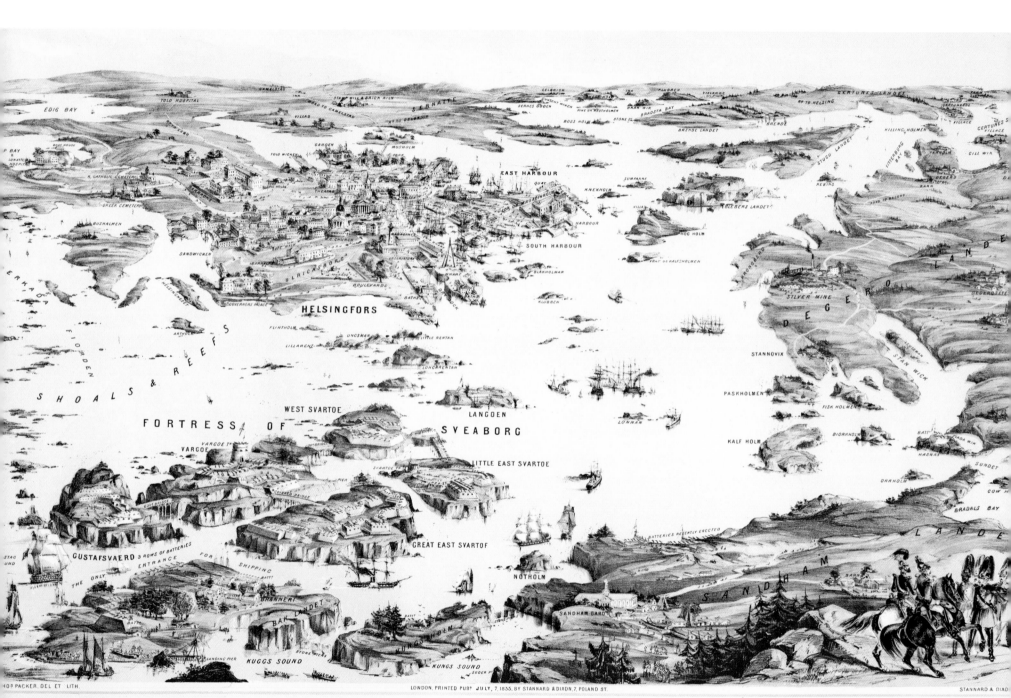

PANORAMIC VIEW OF THE FORTRESS OF SVEABORG, COMMANDING THE ENTRANCE TO THE TOWN & HARBOUR OF HELSINGFORS IN THE GULF OF FINLAND,

N.º 10   THE FORTRESS OF SVEABORG IS BUILT ON SEVEN ROCKY ISLANDS, THROUGH WHICH LIES THE ONLY CHANNEL FOR SHIPPING, THEY MOUNT UPWARDS OF 2000 GUNS, MOST OF WHICH CAN BE BROUGHT TO BEAR UPON VESSELS ENTERING THE HARBOUR. THE ISLANDS ARE CONNECTED BY MEANS OF WOODEN BRIDGES & MOST OF THE BATTERIES ARE BOMB PROOF, THE HARBOUR & TOWN OF HELSINGFORS ARE THE GREAT COMMERCIAL DEPÔT OF THE TRADE & COMMERCE OF NORTHERN RUSSIA.

PROJECTED FROM ILLUMINATED CHARTS & IMPERIAL RUSSIAN SURVEYS RECENTLY PUBLISHED AT S.T PETERSBURG.

triangulation of India in 1800, and became Superintendent of the Great Trigonometrical Survey of India, established in 1817 and largely completed in 1890. Colin Mackenzie became first Surveyor-General of India in 1819. The survey was a formidable achievement, one comparable to the development of an Indian rail system and one unprecedented under India's earlier rulers. An aspect of control was that brought by naming positions in maps.

Beyond the bounds of imperial control, it was necessary to rely on secret agents. Thus, from 1865, the British sent Indians, trained as surveyors but disguised as pilgrims or traders, to Tibet. They were provided with concealed instruments as well as strings of prayer beads containing eight fewer than the usual 108 beads in order to help establish distances. The first of these 'pundits' reached Lhasa, the capital, in 1866, and their information led to the mapping of the major routes, although not of surrounding regions. This mapping greatly helped the British when they sent an expedition that captured Lhasa in 1904, the first from the direction of India.

Because mapping was associated with imperial conquest and control, it was frequently highly unwelcome. In New Zealand, the Maoris often resisted British surveying, and as a result violent clashes occurred in 1843, 1863 and the 1870s. A triangulation survey of the country began in 1875, but again with Maori opposition.

In the United States, the institutional framework for mapping was provided by the army's Corps of Topographical Engineers, which was created in 1838 when the Topographical Bureau was expanded and organised under army control. Force was a powerful factor in this mapping. In 1849, James Simpson's exploration and mapping of what was to be north-western New Mexico and north-eastern Arizona was

undertaken as part of John Washington's punitive expedition against the Navajo. Simpson was able to explore the Canyon de Chelly only after Washington had defeated them. This pattern was more generally the case in the American West.

In Britain, alongside the existing institutions and posts of the Ordnance Survey and the Hydrographer to the Navy, the Depot of Military Knowledge was created in 1803 within the Quartermaster General's Office at Horse Guards, London. This office continued to generate material under its own name until 1857. As an instance of the range and intensity of imperial military activity, maps were also printed by the Quartermaster General's Office at Dublin Castle, by the relevant office at Colombo (Sri Lanka), where the British had established a Ceylon Survey Department in 1800 soon after conquering the coastal regions from the Dutch, and, later, by the Quartermaster General's Department in Simla in India.

As so often, war created new demands for maps and also revealed serious deficiencies in the existing provision. This was very much the case with the Crimean War (1854–56), in which Britain found Russia an unexpectedly difficult opponent in an area about which the British knew little. Crimea was a British target because, at Sevastopol, it contained Russia's Black Sea naval base and, from there, Turkey's position could be challenged, as with the crushing Turkish naval defeat off Sinope in northern Turkey in 1853.

Although the position was much better than when Britain had prepared, without maps, for war with Russia in the Ochakov Crisis of 1791, there was still a shortage of the required geographic information, let alone the maps that best conveyed it.

As a result of these deficiencies, the Topographical

BALTIC OPERATIONS DURING THE CRIMEAN WAR. PUBLISHED BY STANNARED AND DIXON, 1855 A panoramic view, not to scale, that dramatised Baltic operations during the Crimean War. Whereas in 1854 the British fleet achieved little in the Baltic, in 1855 it successfully bombarded Sveaborg, the fort that guarded the approach to Helsinki (then ruled by Russia), destroying not only fortifications but also six ships of the line and seventeen smaller warships sheltering in the harbour. The British made valuable use of sixteen mortar vessels and sixteen screw gunboats which presented a smaller target to the defending batteries and came close inshore. LEFT

THE BATTLE OF THE ALMA, 20 SEPTEMBER, 1854

Having landed on the western coast of
Crimea north of their target, the Russian
Black Sea naval base of Sevastopol, the
Anglo-French forces marched south,
facing the numerically-stronger Russians
on the other side of the River Alma. In
the end, more effective firepower broke
down the strong defence and ensured
that the frontal attack was successful.
The poorly-trained Russians suffered
heavily from the absence of rifles and
rifled artillery. Had there been no such
capability gap, the frontal attack would
certainly have led to very heavy
casualties on the Allied side, but the
Allied combination of percussion-lock
rifle and Minié bullet was deadly. As a
result, the effective range of infantry
firepower increased and the casualty
rates inflicted on close-packed infantry
rose dramatically. **RIGHT**

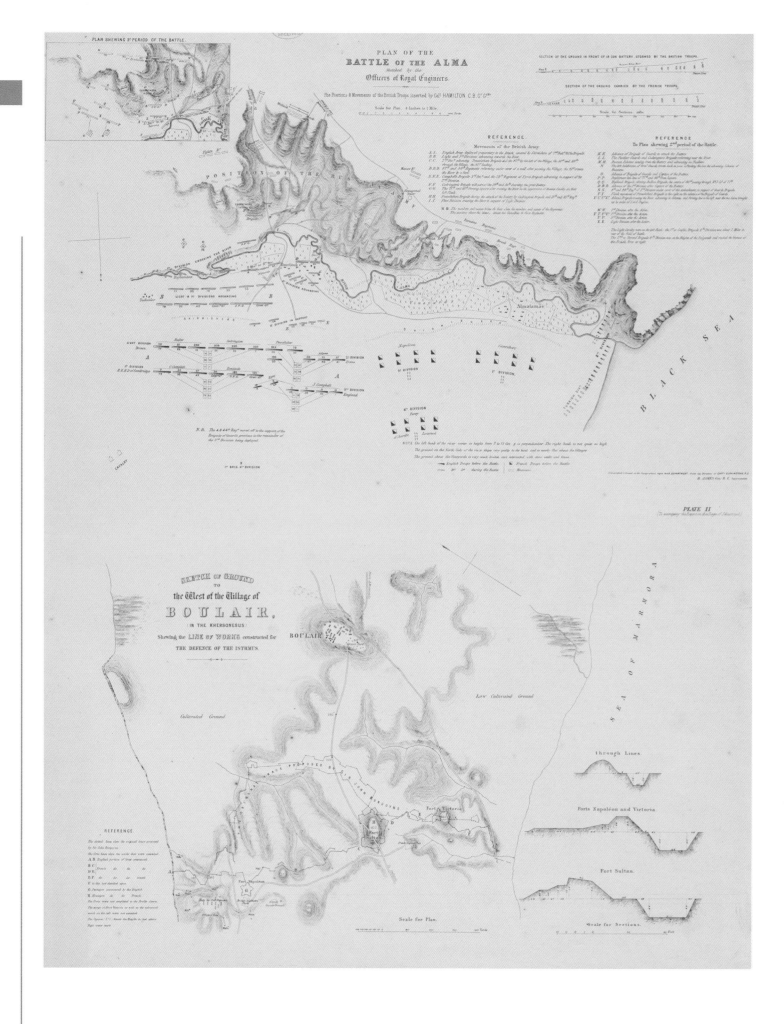

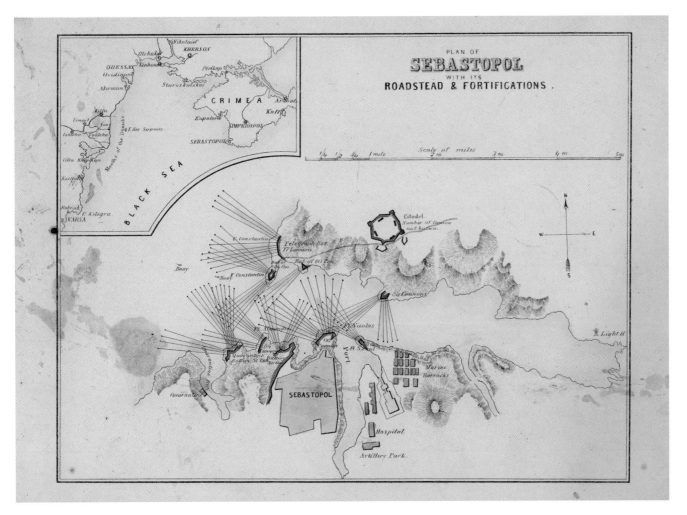

SIEGE WORKS AT SEVASTOPOL, 1854 Despite heavy artillery support, Anglo-French land assaults on Sevastopol in 1855 initially failed, for the city was well defended. Moreover, the attacking Allied armies lacked adequate experience in siege craft and had to face a type of trench warfare. The Russian army was strongly entrenched outside the town, making able use of earth defences, and was supported by more than 1,000 cannon. The scale of warfare was indicated by the Allies firing 1,350,000 rounds of artillery ammunition during the siege, while the impact of the fire on the senses was conveyed by Thomas Molyneux Graves from outside Sevastopol in 1854: 'A few days will show an immense number of guns in battery and when our fire opens it will be tremendous ... now I am so accustomed to the noise that I believe I could go to sleep in a battery when the enemy were firing at it.' An observer added: 'Heavy guns are pouring their dull broadsides on our straining ears.' LEFT

and Statistical Depot was created in London in 1855, and in 1857 this depot and the Ordnance Survey were brought together under the War Office. By 1881, the key cartographic agency in London was the Intelligence Department of the War Office, although maps were also produced by the Royal Engineers' School of Military Engineering, the Royal Engineers Institute, the Directorate General of Fortifications and the Barrack Office.

Alongside the role of government came public interest, which was very much driven by international competition and war. Thus, the British followed Russian advances in Central Asia on maps of the 'Northwest Frontier' of British India, now the north-west frontier of Pakistan. In doing so, however, and noting the extent to which the Russians were advancing, the British unduly minimised the serious problems the Russians faced with the terrain and the native population. In December 1838, the Duke of

Wellington, the victor of Waterloo, wrote to the leading London bookseller, John Hatchard, requesting a copy of 'Arrowsmith's map of Central Asia'. John Arrowsmith (1790–1873) was an active map-maker who published a world atlas in 1834 and was active in the Royal Geographical Society. Arrowsmith's maps drew on the latest geographical information, that provided by commanders as well as explorers.

In 1877, as British anxieties about Russian expansion in Europe and Asia reached new heights, Robert, 3rd Marquess of Salisbury, the Secretary of State for India and later Prime Minister, declared in Parliament:

I cannot help thinking that in discussions of this kind, a great deal of misapprehension arises from the popular use of maps on a small scale. As with such maps you are able to put a thumb on India and a finger on Russia, some persons at once think that the political situation is alarming

INDIAN MUTINY IN DELHI, BY CAPTAIN F.C.

MAISEY, 1857 This map of the Fort and
Cantonment (on a ridge outside the city)
of Delhi shows the British positions on 1
August 1857. It was reduced from the
official map and filled in by Captain F.C.
Maisey. At the outset of the mutiny, Delhi
fell to the rebels on 11 May. The British
force arrived on 7 June, encamping on the
low ridge near the city, but, initially,
lacked the siege guns to attack the city.
From 11 September, a heavy
bombardment began, and, three days
later British storming parties attacked,
finally, after bitter street fighting, taking
control on 20 September. RIGHT

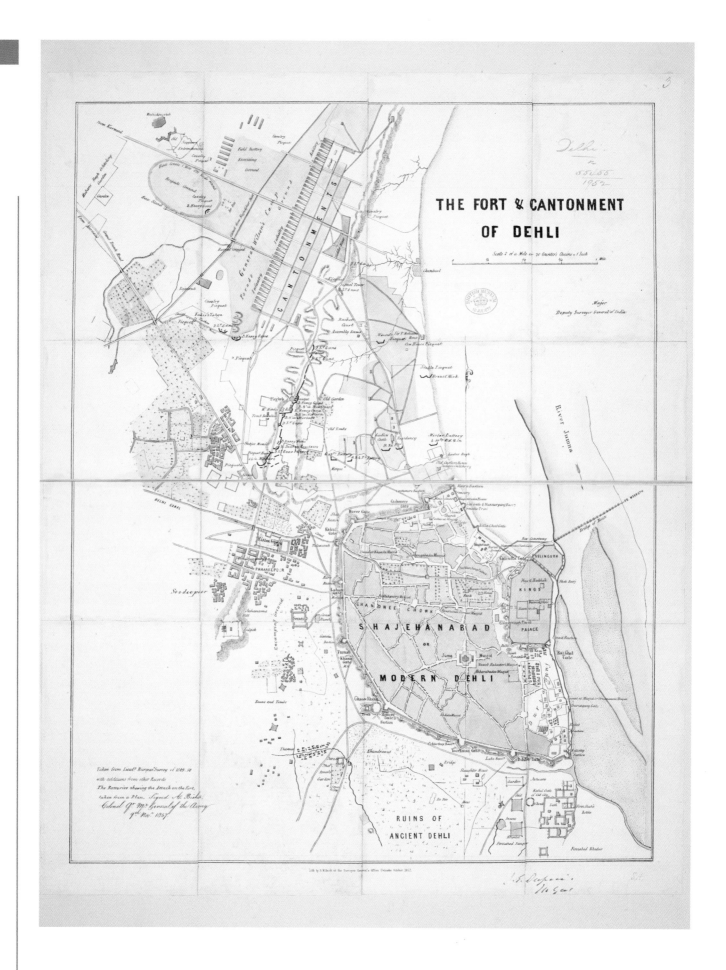

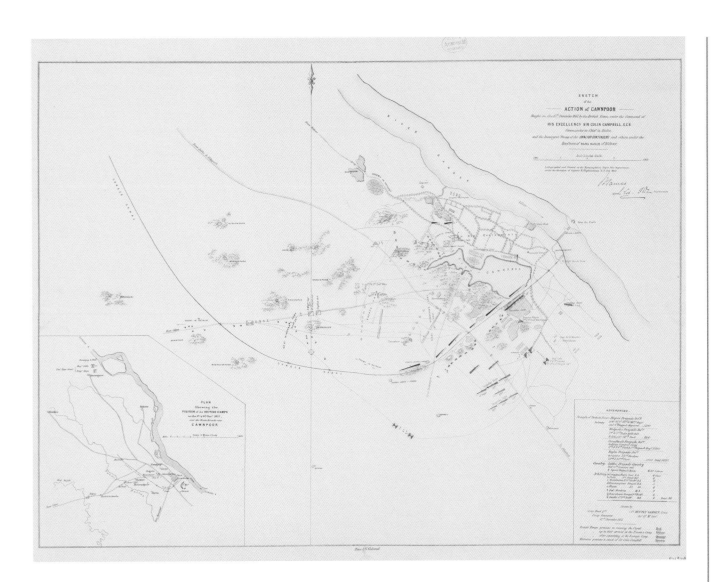

BATTLE AT CAWNPOOR (KANPUR) INDIA, 1857
Military map, printed in the Topographical Depot of the British War Office in February 1858 of the action fought on 6 December 1857. Cawnpoor had been garrisoned by four regiments of native infantry and a British battery of artillery under Major-General Sir Hugh Wheeler. The mutiny there was followed by surrender on terms on 27 June 1857, after which the Indians broke the terms and massacred all the men, women and children. Subsequently Cawnpoor was the seat of much conflict, with Sir Henry Havelock capturing the city, only for Lieutenant-General Sir Charles Windham to be driven from the town in late November. On 6 December, as this map shows, the British Commander-in-Chief in India, Sir Colin Campbell, regained the town. To Victorians, this was an epic as well as moral struggle. **LEFT**

and that India must be looked to. If the noble Lord would use a larger map – say one on the scale of the Ordnance Map of England – he would find that the distance between Russia and British India is not to be measured by the finger and thumb, but by a rule.

No such map was available, but Salisbury, in urging caution about Russian expansionism, was stressing that maps had to be understood if they were to be used

effectively. Arrowsmith had been working at the India Office on a map of Central Asia on the scale of about ninety miles to an inch when he died in 1873. The wide-ranging and multi-faceted 'Eastern Question' led to many references to maps. For example, at a time of Russo-Chinese tension, the *Manchester Courier*, in its issue of 6 November 1880, reported that a Chinese spy had been captured near the Russian Pacific naval base of Vladivostok with a map of the city.

A similar point to that of Salisbury was apparent in other instances, notably during the Cold War when a Soviet zeal to expand in order to seize warm-water ports was readily discerned on small-scale maps. In 1979–89, for example, the Soviet intervention in Afghanistan was depicted in terms of a supposed threat to the Indian Ocean, which, despite the development of a major Soviet air base near Kandahar, certainly did not reflect the military realities in Afghanistan.

REGIONAL INTEREST

War in the nineteenth century, as in other periods, led to interest in other regions. For example, as a result of the Mexican–American War of 1846–48, John Distrunell found that the map of Mexico that he had published in 1846 was in such demand in the United States that he brought out several editions. In this war, Americans wanted to follow operations in California and the West, as well as in what rested with Mexico after the war. These operations required more detailed coverage than that offered by existing maps.

The Crimean War (1853–56) encouraged the production of maps in Britain and her ally France for the public. In 1855, Read and Co. of London published *A panoramic view of the seat of war in the north of Europe*, a map that adopted an aerial perspective on Anglo-Russian hostilities in the eastern Baltic. In this conflict, the British navy failed to lure its Russian counterpart (which remained in St Petersburg) into battle and, instead, staged a number of amphibious attacks and bombardments, notably against the Russian province of Finland. The politics of the war was not readily captured in maps: the threat to St Petersburg was an important factor in precipitating the end of the war and Britain's coastal attack capability helped to leverage peace talks.

As a reminder that national map cultures, in both production and use, were highly significant, there was no comparable appearance of maps in Russia and still less in Turkey, which was involved in the war as an opponent of Russia and ally of Britain and France.

Public interest also led to the production of maps of recent wars. *Maps and plans, showing the principal movements, battles and sieges, in which the British army was engaged during the war from 1808 to 1814 in the Spanish peninsula and the south of France* (London, 1840) by James Wyld was followed by the excellent atlas that accompanied the *History of the war in France and Belgium in 1815* (London, 1844) by William Siborne. Wyld's work was clearly patriotic. It was dedicated 'to the British Army as a tribute humbly offered to its meritorious services and its high character', and in the key the French were referred to as 'the Enemy'.

In Germany, *Die Elemente der Militär-Geographie von Europa* (Weimar, 1821) by Friedrich Wilhelm Benicken, a retired captain in the Prussian army, was followed by the *Atlas des plus mémorables batailles, combats et sieges* (Dessau, 1847) by the Freiherr von Kausler, a major-general in the Württemberg army. At the same time, Siborne's questioning of officers who had taken part in Waterloo revealed marked discrepancies in accounts of the location of units and the timing of action. This is more generally instructive when maps of battles are considered.

War as the cause of national greatness emerged repeatedly in atlases. Based on the work of Karl von Spruner, a Bavarian army officer, the *Spruner-Menke Hand-Atlas für die Geschichte des Mittelalters und der neueren Zeit* (Gotha, 1880) proved particularly influential in Germany and stressed military themes, including a big map of the war zone of the Franco-Prussian War of 1870–71. Johann Droysen's *Allgemeiner historischer Handatlas* (Leipzig, 1886) included a detailed map of the victory at Sedan in 1870 when Napoleon

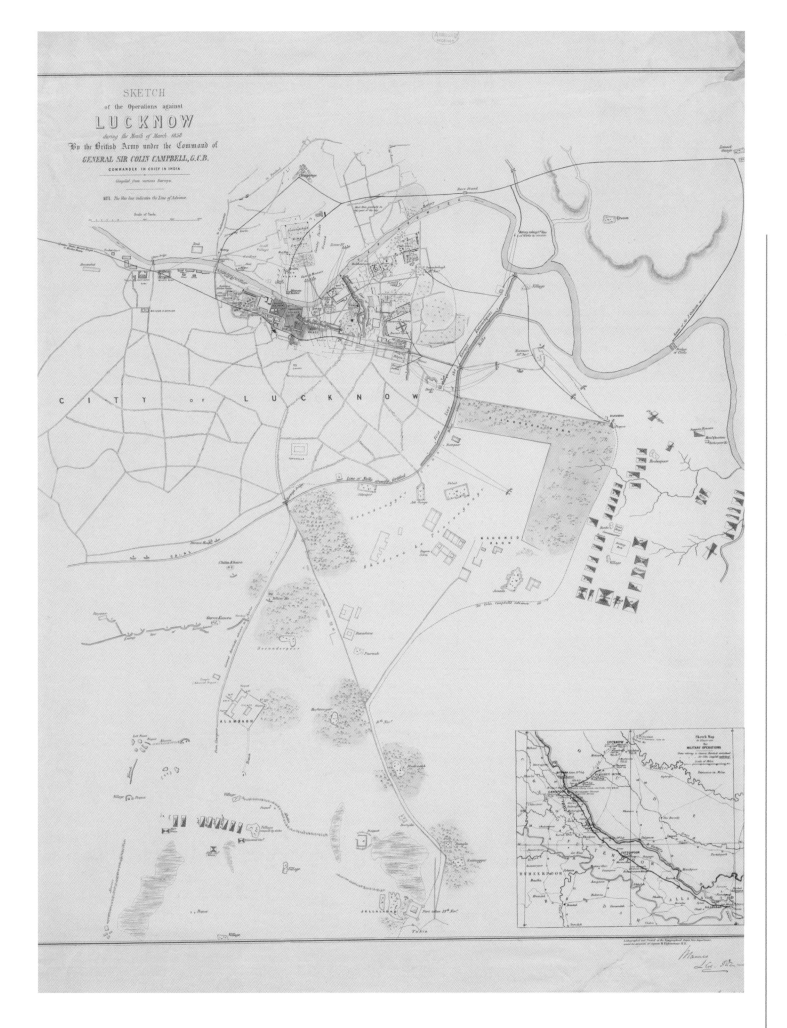

SKETCH
of the Operations against
LUCKNOW
during the Month of March 1858
By the British Army under the Command of
GENERAL SIR COLIN CAMPBELL, G.C.B.
COMMANDER IN CHIEF IN INDIA

THE OPERATIONS AGAINST LUCKNOW IN MARCH 1858 BY THE BRITISH ARMY Lucknow, capital of the state of Oudh, had been besieged by mutineers from June to November, being relieved by the British Commander-in-Chief in India, Sir Colin Campbell, in November 1857. He then evacuated the city. In March 1858, Campbell finally recaptured the city after ten days' hard fighting. LEFT

CAPTURE OF CANTON, MAP BY FRENCH ARMY, 1857

This map, produced by the French army, distinguished the operations of Anglo-French forces on two crucial days during the Second Opium War. Meanwhile, the crisis of the long-lasting Taiping rebellion lessened any chance that the Chinese would be able to resist Anglo-French pressure. Initially in this war, the British relied on a naval response, seizing the forts on the approach to Canton, but the need to divert troops to deal with the Indian Mutiny made it difficult to take matters further until French reinforcements arrived. In 1858, attention shifted to north China and an attempt to put pressure on Beijing which was captured in 1860. RIGHT

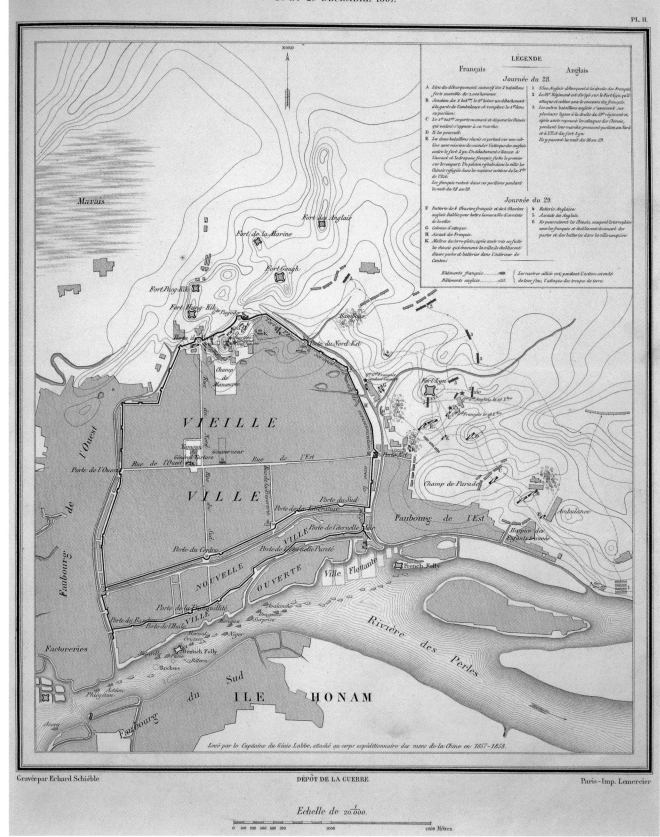

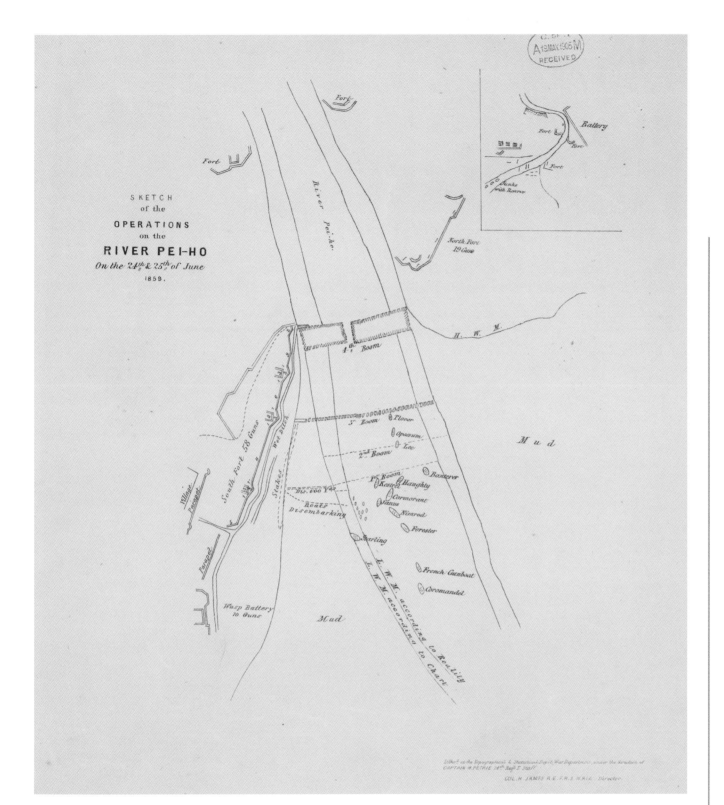

SKETCH
of the
OPERATIONS
on the
RIVER PEI-HO
On the 24ᵗʰ & 25ᵗʰ of June
1859.

SKETCH OF THE OPERATIONS ON THE RIVER PEI-HO ON 24–25 JUNE 1859, PUBLISHED BY THE TOPOGRAPHICAL AND STATISTICAL DEPOT OF THE BRITISH WAR DEPARTMENT A British fleet under Admiral Sir James Hope attempted to force a passage with eleven gunboats and a landing force of 1,000 men, but met severe resistance. Six of the gunboats were sunk or disabled and the British lost 89 killed and 345 wounded. The British found Chinese fire as accurate as their own. Owing to the shallows, the British ships of the line and frigates had to stand off out of range of the forts at Taku, leaving the wooden screw gunboats to take the Chinese fire. The earthen Chinese fortifications absorbed British shot. In 1860, however, a fresh attack by a larger Anglo-French force led to the fall of the Taku forts. **LEFT**

III and 83,000 troops surrendered on 2 September after the failure the previous day to drive the Prussians and their artillery from the surrounding hills; 21,000 men had already been captured in the battle. Prussia's victorious leadership of Germany emerged clearly from such maps and atlases, and from subsequent maps of Germany for the war ended with the annexation of most of Alsace and much of Lorraine.

More generally, the military played a major role in the production of maps and atlases. The *Atlas physique, politique et historique de l'Europe* (Paris, 1829) by Maxime-Auguste Denaix was published by a graduate of the *École Polytechnique* who had gone into the army before moving to the *Dépôt Général de la Guerre*. It was engraved by Richard Wahl, who had been trained at the *Dépôt Général*, and published under the sponsorship

BATTLE OF TOURANE (DANANG), VIETNAMESE

MAP, 1859 In 1857, the execution of two Spanish missionaries by the Vietnamese Emperor Tu Duc provided an excuse for French intervention in Vietnam. In 1858, a joint French and Spanish expeditionary force landed at Danang. In turn, Saigon was seized in 1859. The Vietnamese had borrowed Western models for the uniform, arms and discipline of the army, as well as fortifications, but it proved difficult for them to translate form into substance. Tactically and technologically, the Vietnamese did not match leading-edge Western developments and, more seriously, they did not choose to embrace the process of continuous change which was increasingly potent in the West. This choice owed something to cultural preferences, especially to the conservative nature of Confucianism and to its role in education. But it owed even more to tasking: the army was largely designed to cope with peasant uprisings.

RIGHT

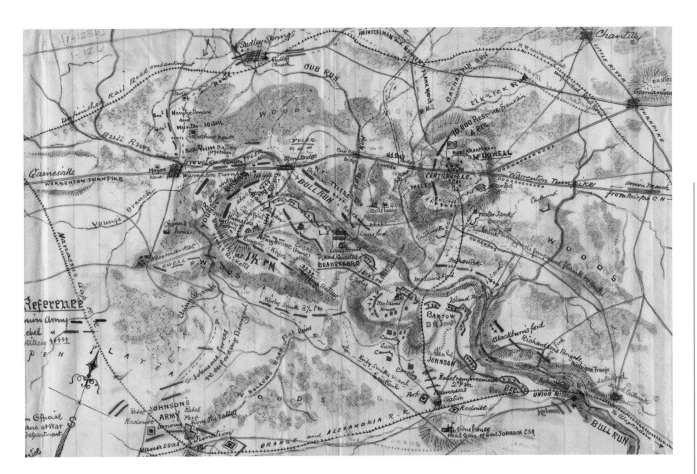

FIRST BATTLE OF BULL RUN, 1861. BY ROBERT KNOX SNEDON Any hopes that the American Civil War would be rapidly brought to a close were ended by failure as neither commander behaved adroitly. The fate of the battle on 21 July 1861 hinged not on planning but on the arrival of fresh troops. In that, the Confederates benefited from operating on interior lines, and their reinforcements from the Shenandoah Valley decided the battle with an attack on the Union flank. The battle ended with the flight of the Union forces, who suffered heavier losses, although the Confederates failed to exploit their victory, in large part because they had been exhausted and disorganised by the fighting. **LEFT**

of the Vicomte de Caux, the Minister of War. By the time of the 1836 edition, which was published with the approval of Caux's successor as Minister of War, Denaix was head of the administration at the *Dépôt*. An important continuity in expertise had developed there, and there was to be a comparable continuity elsewhere, one that looked toward the creation of General Staff systems.

TECHNOLOGICAL ADVANCES

Technological changes were of great significance in encouraging map-making. In the 1800s, mechanised paper-making became commercially viable, leading to the steam-powered production of plentiful quantities of inexpensive paper. The steam-powered printing press developed in the same period, transforming the economics and practicalities of scale.

Moreover, map-colouring ceased to be a manual process and was transformed by the onset of common colour printing. Colour printing made it less expensive to convey more information. The density and complexity of information that could be conveyed increased, as with colour-coded contour zones.

Lithography made a major impact in the 1820s, and was less expensive than engraving. Moreover, lithography itself developed as metal plates were substituted for stone and the transfer technique was developed. This led to a much finer line and to neater lettering.

GENERAL STAFF SYSTEM

In a military world in which planning, and personnel specifically for planning, came to play a greater role, maps became more important. They were the fundamental tool of the General Staff system and, as it rose in significance, so planning and mapping became more important. Maps were important for a systematic process of effective and rapid decision-making and for the implementation of strategic plans in terms of timed operational decisions and interrelated tactical actions.

The land survey section of the Prussian General Staff, which was established in 1816, was responsible for producing maps, and it trained its officers in the

Winfield Scott, the General-in-Chief and a distinguished veteran, called for an advance down the Mississippi, to bisect the Confederacy, combined with a naval blockade. Termed the 'Anaconda Plan' by the press, this was intended to save lives and, by increasing support for a return to the Union, to encourage the Confederacy to make peace or, failing that, to put the Union in the best state for further operations. However, Scott's emphasis on planning and preparation, as well as on the indirect approach, training and a delay in the offensive until the autumn of 1861, fell foul of the pressure for action. Scott sought what he termed 'a war of large bodies' and, instead, 'a little war by piece-meal' prevailed, notably with the advance that led to what became the First Battle of Bull Run. RIGHT

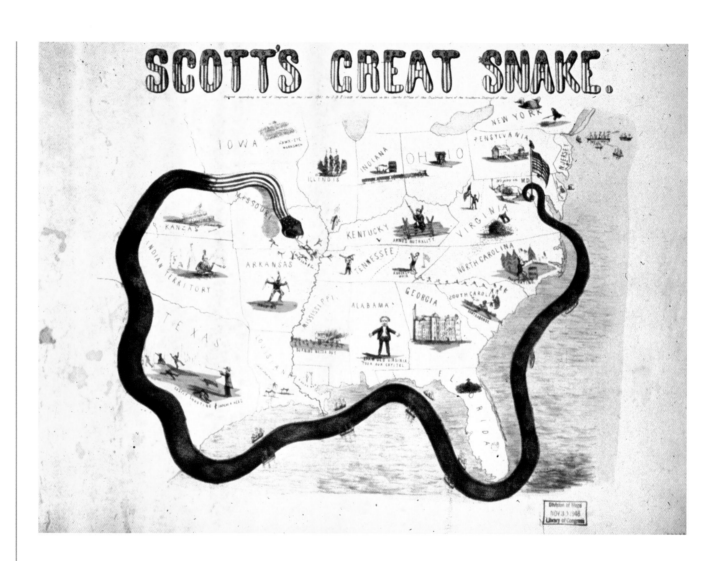

SCOTT'S GREAT SNAKE.

necessary trigonometrical, topographical and cartographical skills. All officers in the General Staff were expected to work in the section for a number of years. Four of the six heads of the General Staff between 1820 and 1914 spent several years there. Indeed, the most successful, Helmuth von Moltke the Elder, the commander during the Wars of German Unification in 1866–71, spent his first three probationary years in the section. Its maps were important to the war gaming and manoeuvres seen as crucial to the staff officers' education. Moltke was

particularly aware of the role and potential of railways, a new means of operation very much linked to maps.

Other states sought to match Prussia's methods after its spectacular victories over Austria and France in 1866 and 1870–71 respectively. During the latter, the French army had had few maps of eastern France, and the maps were not as good as those of their German opponents, but this changed as Prussian staff work was copied. The *École Supérieure de la Guerre* was founded in 1878 in order to provide France with a staff college, and the Army Geographical Service followed in 1887.

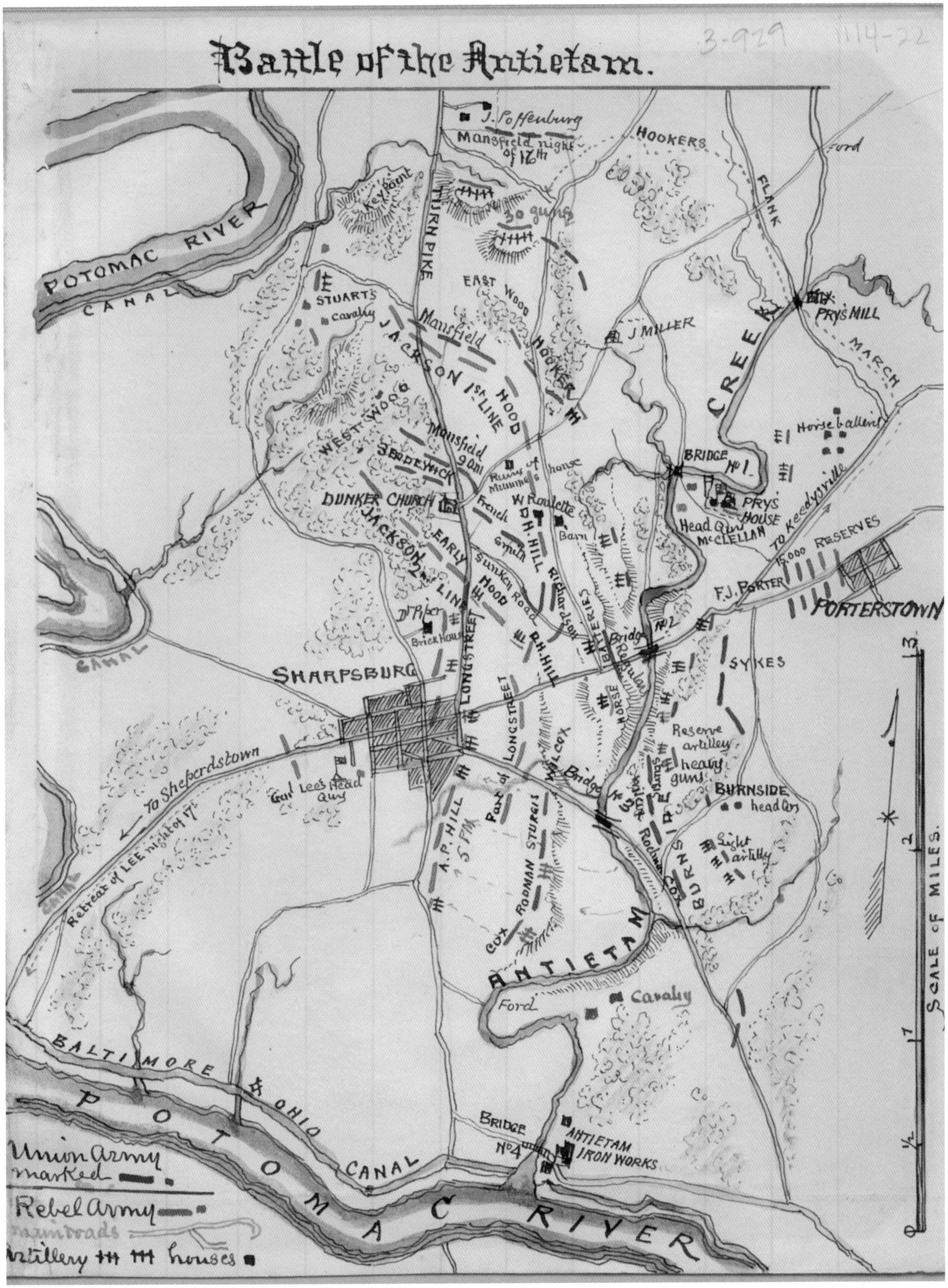

Battle of the Antietam.

BATTLE OF ANTIETAM, 17 SEPTEMBER, 1862. BY ROBERT KNOX SNEDEN George B McClellan, the lacklustre commander of the Army of the Potomac, the leading Union army, failed to implement his plan for hitting the Confederate flanks before breaking the centre. Instead, a series of piecemeal attacks on 17 September 1862 was unable to provide mutual support, and this exacerbated their costly character as frontal assaults. The attack on the Confederate left was mounted before that on the centre, let alone the right. The eventual Union breakthrough on the Confederate right, on Antietam Creek, came too late to determine the flow of the battle, and McClellan failed to develop his victory. This failure helped the outnumbered Confederates to fight to a draw and to stand their ground before withdrawing two days later. The Confederate defenders had taken heavy casualties because they were not entrenched. **LEFT**

VICKSBURG, 1863. BY ROBERT KNOX SNEDON As
Union hopes of a quick war faded, the
emphasis came to be on how best to win
a long conflict. This led to a focus on
securing control of the Mississippi River
and splitting the Confederacy, as the first
stage of operations. This focus interacted
with the opportunities created there by
Union success. Vicksburg was the crucial
Confederate fortress on the river. Once
he had made the campaign there fluid,
Ulysses S Grant, the Union commander,
gained opportunities to achieve
concentrations of strength that enabled
him to take successful initiatives and
drive the Confederates in on the city
where they could be besieged. The
American envoy in Paris claimed: 'The loss
of this river was more injurious to the
cause of the insurrectionists than the loss
of many battles.' The ruthless Grant lived
off the land, which helped him deal with
supply issues. Artillery, however, was not
available in the numbers seen to be
necessary in the 1910s. The Union
batteries mounted only 220 guns over
twelve miles of siege line in June 1863.
RIGHT

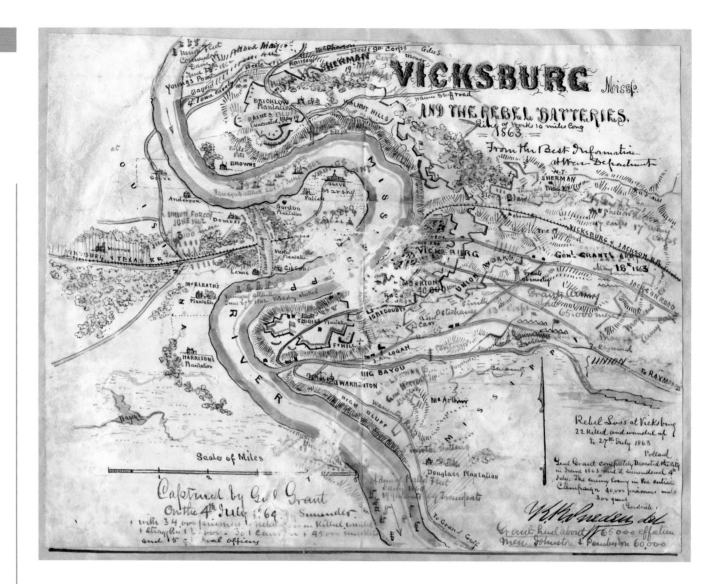

It had sections on geodesy, precision surveys,
topography, cartography, the construction of relief
models and precise instrumentation, and was under the
General Staff.

Other states followed suit. In Argentina, the
foundation of the *Colegio Militar de la Nación* in 1870
and the *Escuela Naval Militar* in 1872 was followed in
1900 by that of the *Escuela Superior de Guerra*, a war
college for senior officers.

In many countries, the standard medium-scale
topographical map was known as the General Staff
map. Indeed, the General Staff surveying office was
responsible for 545 sheets of a map of Germany at the
scale of 1:100,000. This model was not restricted to
Europe. Prussian influence led the Turkish General
Staff to create a cartographic section in 1895. Seven
years earlier, the Japanese army had founded the

Japanese Imperial Land Survey within the General
Staff, but the Chinese Military Survey Institute did
not follow until 1902.

Consideration as to how best to present cartographic
information to help military planning and operations
led to a discussion about how much information to
convey. The influential Committee on the Military
Map of the United Kingdom reported in 1892 in
favour of coloured maps with contours, road
classification and the use of symbols, not names, in
order to achieve less clutter. These methods became
more common as the goal of useful, uniform
maps was pursued.

AMERICAN CIVIL WAR

The peacetime response to the need for maps seen in
Europe in the last three decades of the century, not

BIRDS EYE VIEW
OF THE
SEAT OF WAR AROUND RICHMOND

BIRD'S-EYE VIEW OF THE SEAT OF WAR AROUND RICHMOND, BY JOHN BACHMANN, 1862 This map shows the Battle of Chickahominy River on 29 June 1862. John Bachmann's aerial view demonstrates the challenge to the Confederate capital, Richmond, after a large-scale Union amphibious landing to the east of the city. The Union campaign initially seemed promising, with the Confederate forces retreating in the face of their numerous but ponderous opponents. However, taking over from Joseph Johnson after he was wounded at Fair Oaks on 31 May, Robert E Lee succeeded in blockading George McClellan's cautious advance in the Seven Days battles (25 June – 1 July 1862). These battles began a series of Southern advances and victories in the east. **LEFT**

least in terms of the frequent training manoeuvres of conscripts and reservists, had been matched in the United States by a somewhat different wartime response to crisis during the Civil War (1861–65), which created issues for both the military and the public. At the outset, the field commanders on both sides were seriously hampered by a shortage of adequate maps. They needed to use maps for strategic, operational and tactical reasons. Maps were central to

devising overall strategy. The scale of operations, and the need to coordinate and move forces over considerable distances, within and between individual campaigns, both encouraged the use of maps, as did the unfamiliarity of much of the terrain and the need for detail.

Aside from at the level of campaigns, the scale of battles was such that it was no longer sufficient, tactically, to rely completely on the field of vision of an

BATTLE OF GETTYSBURG, 1–3 JULY, 1863.

PUBLISHED BY C. A. ALVORD, ELECTROTYPER AND

PRINTER, NEW YORK, 1863 A lack of adequate reconnaissance led Robert E Lee, the commander of the 'rebel' (the Confederate Army of Northern Virginia) into battle at Gettysburg on 1–3 July 1863 without clear knowledge of what was in front of him. Lee failed to coordinate his attacks on the second day, which allowed George Meade, the Union commander, to benefit from his interior position in countering them. The movement of Longstreet's Texan and Alabama regiments to the flank was readily observed by Union commanders who were able to move Chamberlain's regiment to Little Round Top in order to counter the flanking threat. On the third day, the Confederate attack on the Union centre, Pickett's Charge, did not receive adequate flank support. Lee retreated after the battle and this, combined with Grant's success at Vicksburg on the Mississippi, which surrendered on 4 July, seemed to suggest that a major shift in the balance of advantage had occurred.

RIGHT

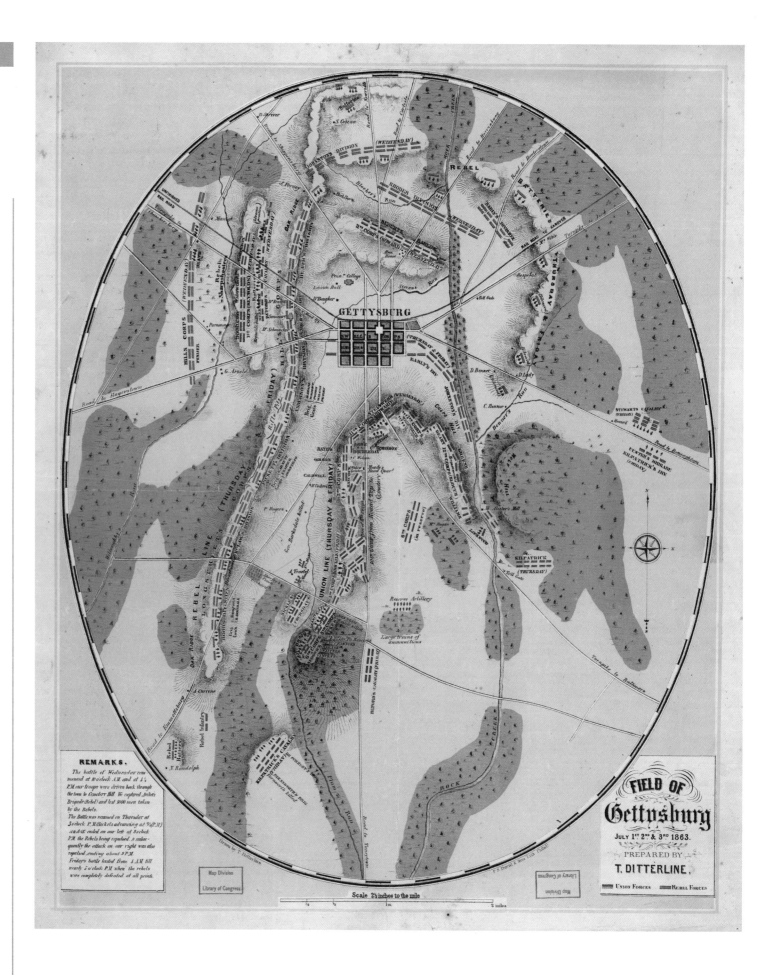

individual commander and his ability to send instructions during the course of the engagement. Instead, it was necessary to plan far more in advance. This was particularly important in preparing artillery positions, in mounting and responding to frontal attacks, and in coordinating attacks from a number of directions.

It was necessary to understand terrain in order to control it, and also for specific reasons, including the interplay of rail links and field operations, and the best use of water routes, such as the Mississippi. The interplay of rail links and field operations was very important in many battles, including First and Second Manassas (also known as Bull Run, 1861, 1862) and Shiloh (1862), whether in feeding troops into the battle or in explaining why particular sites were contested. The Union forces had a particular need for maps as they were advancing south into the Confederacy and seeking to make best use of their more numerous reserves.

Commercial cartography could not produce sufficient maps. As a result, the Confederate and Union armies turned to establishing their own map supplies. By 1864, the Coast Survey and the Corps of Engineers were providing about 43,000 printed maps annually for the Union's army. In that year, the Coast Survey produced a uniform ten-mile-to-the-inch base map of most of the Confederacy east of the Mississippi River. Lithographic presses produced multiple copies of maps rapidly. The production of standard copies was crucial, given the scale of operations. Surveyors and cartographers were recruited for the military, creating problems for private producers.

American newspapers found it appropriate and necessary to produce many maps. They were able to do so because of the presence of military correspondents and recent advances in production technology in printing and engraving. The development of illustrated journalism paved the way for the frequent use of maps. Military correspondents sent eyewitness sketches that were rapidly redrawn, engraved and printed. In addition, publishers issued a large number of sheet maps. The scale of production was vast. Between 1 April 1861 and 30 April 1865, the daily press in the North printed 2,045 different maps relating to the war.

The public need for maps was also served by the publication of large numbers of freestanding maps, especially in the North. These included theatre of war and battle maps, such as that of the vital Potomac area, for example *Bacon's new army map of the seat of war in Virginia* (1862). That year, while Union forces campaigned along the Mississippi, in fulfilment of their ultimately successful aspiration to cut the Confederacy in half along that axis, James Lloyd brought out his *Map of the lower Mississippi River from St Louis to the Gulf of Mexico*. Arresting perspectives were adopted, as by John Bachmann in his *Bird's-eye view of Louisiana, Mississippi, Alabama and part of Florida* (1862), a map that helped to explain the naval campaign of 1862, which led to the capture of New Orleans and Baton Rouge by Union Admiral David Farragut. Bachmann produced a series of striking bird's-eye views of spheres of conflict.

These maps enabled those on the 'home front' to follow the movement of relatives and others in the military. They had a visual impact comparable to the maps of Richard Edes Harrison during the Second World War, again maps with a striking aerial perspective, albeit ones that covered a greater geographical area.

Conversely, in the far poorer Confederacy, there was a severe shortage of press operators, printers, wood-engravers and printing materials, all of which

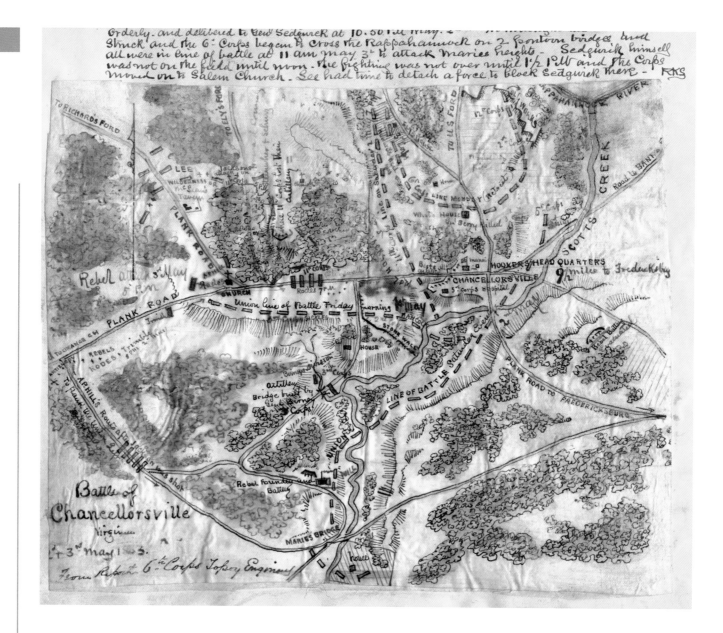

BATTLE OF CHANCELLORSVILLE, 1863. BY ROBERT KNOX SNEDEN This was one of a gruelling series of battles in the east. Joseph Hooker, the Union commander, planned to outflank Lee's left, but initial operational success was lost on 1 May 1863 as Hooker responded cautiously to engagement with the Confederates. Instead, it was Stonewall Jackson's Second Corps that made the most successful flank move, turning the Union right on 2 May before launching an effective attack. Jackson, however, was mistakenly shot by a Confederate soldier in the dark. After a subsequent Union advance was checked on 4 May, Hooker retreated to the other bank of the Rappahannock River, proving that superiority in men and materiel could not yet be translated into an effective army capable of defeating the Army of Northern Virginia; and, specifically, able to bring Union forces to bear and to outfight their opponents. RIGHT

combined to ensure that the appearance of maps in Southern newspapers was rare. However, some, such as the *Charleston Mercury* and the *Augusta Constitutionalist*, did print a few.

Newspaper publication of maps of the conflict led to government action. On 4 December 1861, the front page of *The New York Times* carried 'The national lines before Washington: a map exhibiting the defences of the national capital, and positions of the several divisions of the Grand Union army'. The commander, George McClellan, furiously demanded that the paper be punished for assisting the Confederates. The Secretary of War restricted himself to urging the editor to avoid such action in the future. The

following spring, however, the War Department established a voluntary system to prevent journalists with the Army of the Potomac from publishing compromising maps.

Maps also served to make satirical points, as in *Scott's great snake*, a depiction of the 1861 'Anaconda Plan' of General Winfield Scott for the defeat of the Confederacy.

Following the war, public interest led to maps of the conflict. Thus, John Bachelder published a map of the battlefield in his guidebook *Gettysburg: what to see and how to see it* (1873), while for the War of Independence Henry Carrington produced *Battles of the American Revolution … with topographical illustration* (1876).

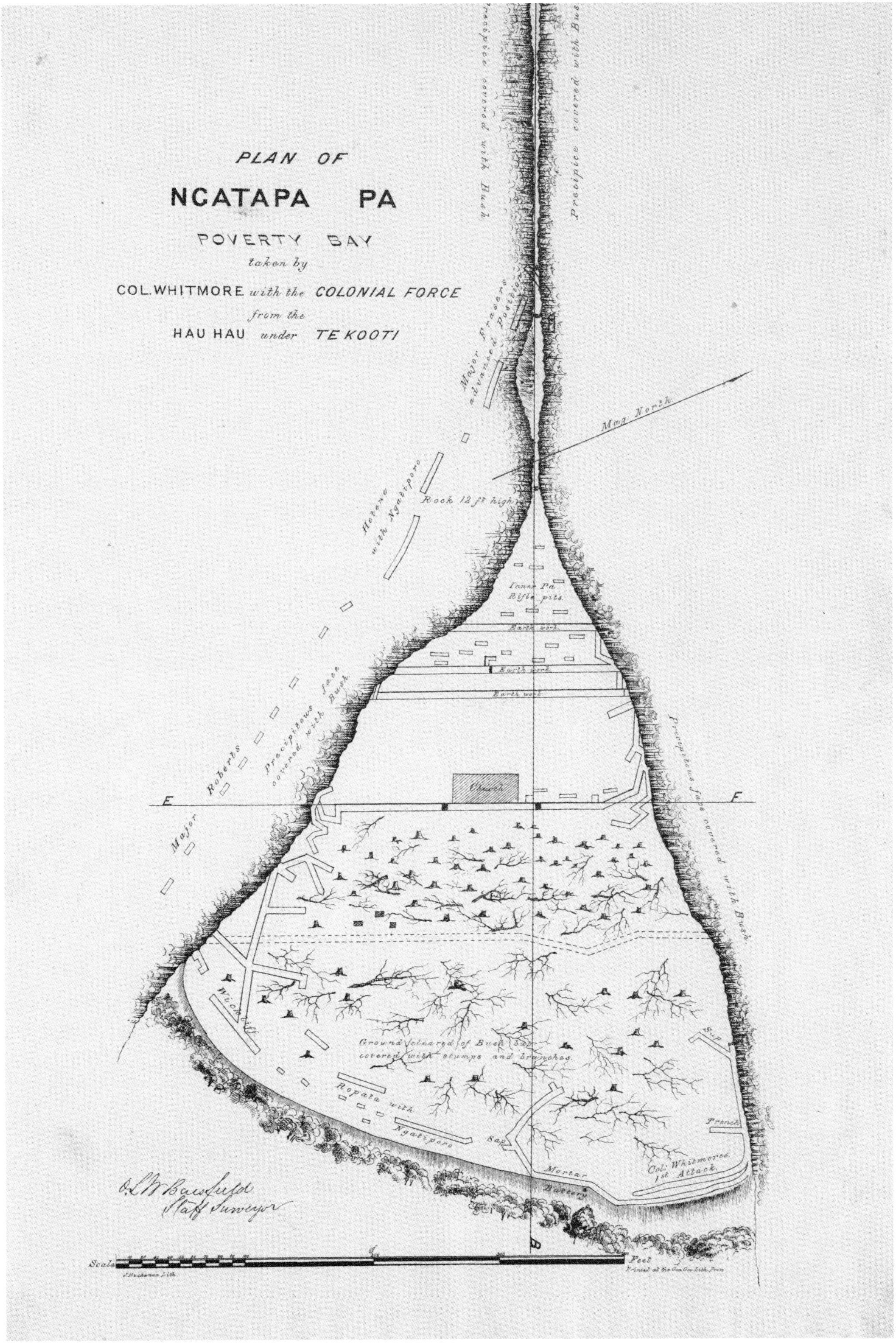

PLAN OF NGATAPA PA

POVERTY BAY

taken by

COL. WHITMORE *with the* COLONIAL FORCE

from the

HAU HAU *under* TE KOOTI

THE MAORI WARS. PLAN OF NGATAPA PA, POVERTY BAY The Maoris, under Titokowaru and, in this case, Te Kooti, used well-sited trench and pa (fort) systems that lessened the impact of British artillery and that were difficult to bombard or storm. The bayonets that were adopted were designed to increase the potential for their own muskets. The Maori inflicted serious checks on the British in the 1860s, but the availability of British, colonial and allied Maori units, and the process of extending control, that included road and fort construction, ensured an eventual settlement on British terms. Although Te Kooti escaped the siege of Ngatapa Pa in January 1869, by fleeing down a steep cliff in the early hours, Maori forces loyal to the government captured and executed more than 130 of his supporters. The pa was a strong ancient hilltop fortress in a good defensive location but, crucially, lacked a secure water supply and the defenders ran out of water. There were about 200 fighting defenders and about 600 attackers. The interior of the pa contained many rifle pits, and it was surrounded by cliffs on three sides. **LEFT**

THE MAORI WARS, NOVEMBER 1863 Map of Rangiriri between Lake Waikare and Waikato River showing location of Maori groups and redoubts and the position taken by the troops with their three Armstrong guns. The Maori defences comprised an entrenched parapet with ditches on both sides, redoubts, and concealed rifle pits protected by wooden stakes. There were insufficient Maori, however, to defend these extensive fortifications from the 860 British troops that arrived overland and by river fleet. On 20 November, the British, who had failed to carry out the necessary reconnaissance, launched frontal assaults, found the bank of the central redoubt too high to scale, and were blocked by tough resistance leading to an inconclusive battle. Each side lost about 50 dead. Overnight most of the Maori withdrew, including the Maori leader. Most of the Maori surrendered the next day. The battle was the key battle in the Waikato campaign. **RIGHT**

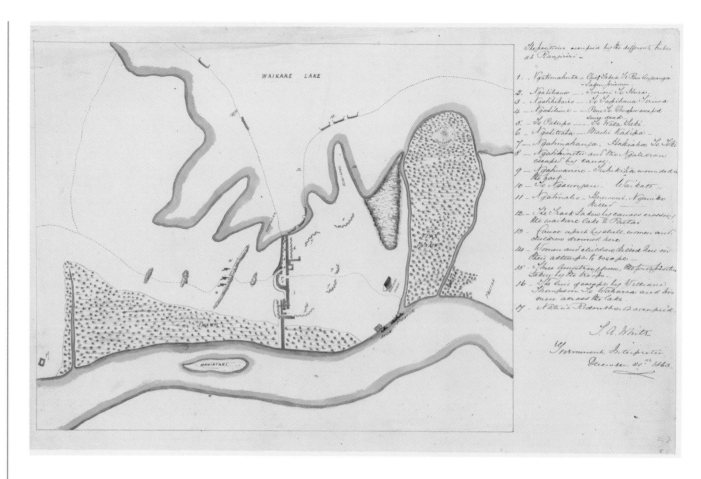

This was part of a more general process of mapping war for a large public, notably a public concerned with news, information and understanding the changing world. This process led to the presentation of areas in terms of war, notably recent conflicts. Atlases of recent wars including the Spanish *Atlas histórico y topográfico de la guerra de Africa en 1859 y 1860* (1861) and the Brazilian *Atlas histórico da guerra do Paraguay* (1871).

The American National Geographic Society first published a map with its magazine in 1899 when America's conquest of the Philippines led to the inclusion of a map of the islands. The war with Spain the previous year (which this conflict stemmed from) had already led to the publication of two atlases in the United States in 1898, the *St Paul Spanish-American War Atlas* and *Shewey's official handy reference pocket and cyclopedia, containing authentic historical information and statistical tables of reference relating to the Spanish-American conflict, with official maps.* Reference to maps and their links with war were already well developed prior to the much more wide-ranging and intense conflicts of the twentieth century.

NATURE OF WAR

Compared with what was to come, the nature of war in the nineteenth century was far easier to map. Balloons played a role from the 1790s, but it was a relatively minor one, and, in essence, the nineteenth

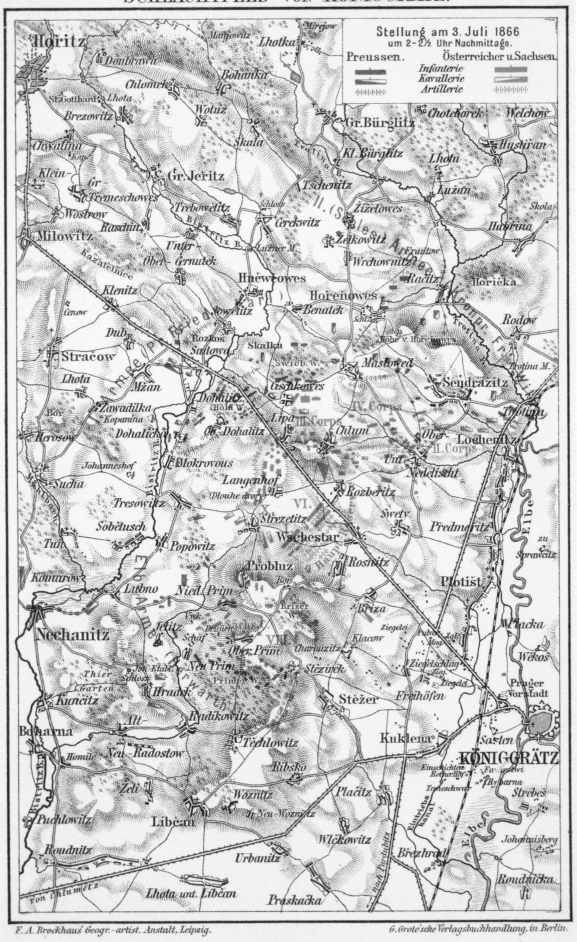

SCHLACHTFELD von KÖNIGGRÄTZ.

Stellung am 3. Juli 1866
um 2–2½ Uhr Nachmittags.

Preussen.   Österreicher u. Sachsen.

Infanterie
Kavallerie
Artillerie

F. A. Brockhaus' Geogr.-artist. Anstalt, Leipzig.   6. Grote'sche Verlagsbuchhandlung, in Berlin.

Maßstab 1 : 140.000.   Kilometer.

BATTLE OF SADOWA/KÖNIGGRÄTZ, 1866 Each
side deployed a quarter of a million men
on 3 July 1866 in what was the decisive
battle in the Austro-Prussian war. The
Austrian commander, Ludwig August von
Benedek, was in a reasonable defensive
position, had better artillery and had the
possibility of using interior lines to defeat
the Prussian armies separately. In
practice, however, due in large part to
Benedek's irresolution and inability to
take advantage of his interior position,
the Austrian situation was less favourable.
His forces also lacked high morale. The
Prussians sought to concentrate their
separate armies in the battle and,
although the Prussian envelopment was
not fully successful, the coherence and
morale of the Austrian army was
destroyed. Moreover, its casualties were
far heavier. At the tactical level, Prussian
units possessed a flexibility their
opponents lacked, ensuring that the
Austrian positions were caught in the
flanks and hit by cross-fire. In contrast,
massed Austrian attacks suffered heavy
losses. **LEFT**

IMPERIALISM AND MAPPING: THE INVASION OF

ABYSSINIA, 1868 Britain invaded Abyssinia
(Ethiopia) in 1868 in order to rescue
British hostages, including the British
Consul, who were held at Magdala by
Emperor Tewodros II (r. 1855–68). An
eccentric, he had tried to modernise the
army in order to strengthen the state and
had imported large numbers of muskets.
In a methodically-planned campaign,
Lieutenant-General Robert Napier led a
large force from India, arrived at Annesley
Bay in the Red Sea, shown as the port of
Malkatto (ie Mulkatto) on the map, and
then advanced into the mountainous
interior. The Ethiopians were defeated at
Arogee, a battle in which the British
outnumbered the Ethiopians by 13,000 to
7,000 men and outshot them with better
artillery and breech-loading rifles,
Magdala was stormed, the captives were
freed, Tewodros committed suicide, his
successor, Yohannes IV, was given arms,
and the British withdrew. Surveyed by
the army in 1868, the map was photo-
zincographed at the Ordnance Survey
Office in 1870. RIGHT

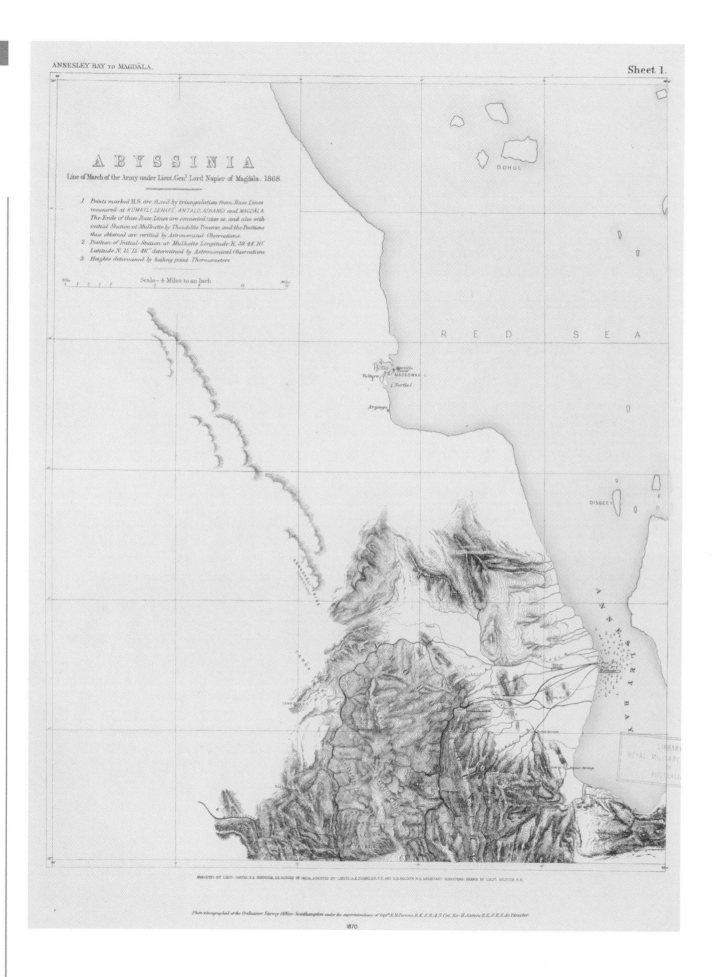

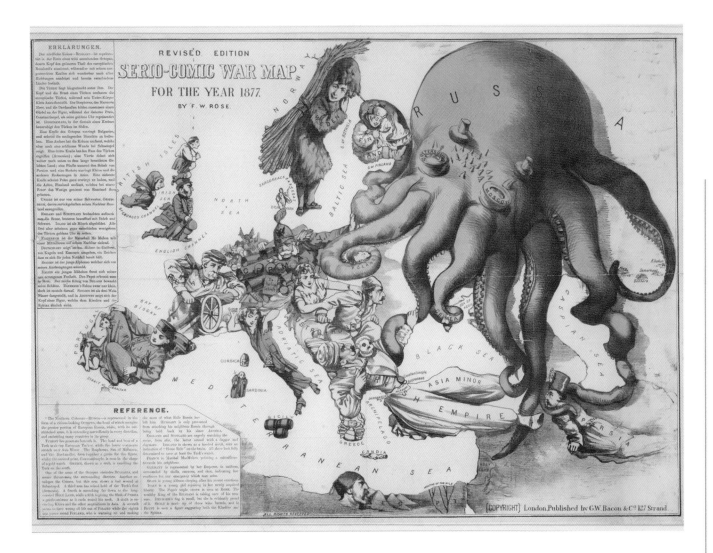

SERIO-COMIC WAR MAP FOR THE YEAR 1877.

PUBLISHED BY G. L. J. BACON & CO, LONDON

Russia is presented in this British map as the expansionist threat. It has already seized Finland and Poland, and is now attacking Turkey and Persia (Iran), advances that challenged British views about the route to India. The Russo-Turkish War of 1877–78 saw the Russians seize the fortress of Plevna (Pleven in modern Bulgaria) after a bitter siege in 1877, and then advance to within fifteen miles of the Turkish capital Constantinople (now Istanbul). British military pressure in 1878 obliged Russia to abandon the ambitious Balkan peace it had imposed on Turkey. **LEFT**

century was the last with a flat battle space, in that the limit of the vertical battle space was that set by the contours and by the height of artillery shot. These factors could be readily understood, and the former was depicted with greater accuracy as surveying improved and contour systems were standardised.

The need to understand the battle space changed greatly during the century with the increasing range and lethality of firearms, both handheld and artillery. A host of developments were significant, including percussion caps, bullets, bolt actions, rifling, breech loading, steel artillery, machine guns and automatic resiting mechanisms. In turn, each of these entailed continuing processes of improvement, frequently involving trial and error.

The net effect was a greater range of weaponry that required maps covering a larger area, and a greater lethality of weaponry that led to less dense troop formations, the so-called 'empty battlefield' and, therefore, again, the coverage of a larger area. For example, rifled artillery permitted greater range, which enabled the artillery to move back out of the range of accurate infantry fire without losing accuracy. The improvement of artillery barrels enhanced their range, accuracy and lifespan. The physical character of the battlefield was affected by the replacement of black-power firearms by those using smokeless bullets. This helped keep the positions from which fire was coming secret and was an element that it was very difficult even to suggest on maps.

So also with fortifications. They were now bombarded from a greater range. As a result, fortifications became more far-flung defensive systems, as with the Russians at Sevastopol during the Crimean War and the Confederates at Vicksburg in 1863. This situation looked toward the construction, later in the

SEAT OF WAR IN ASIA, BY THE OFFICE OF THE CHIEF OF ENGINEERS, 1878 This map of Afghanistan from surveys made by British and Russian officers up to 1875 was published in 1879 by the Office of the Chief of Engineers for the information of the officers of the US Army. British failure in Afghanistan in 1841–42 was followed by invasion in 1878 in an effort to block Russian influence. Prefiguring the situation in much warfare in the later stages of the empire, the British found it easier to advance and hold positions than to end resistance, let alone stabilise the situation. In 1878, a British force advanced on Kabul and defeated the Afghans at Peiwar Kotal. The following May, the Afghans accepted British control over the frontier passes and dominance over Afghan foreign policy, but in September 1879 the newly-installed British Resident and his staff were killed and a fresh advance proved necessary. The British entered Kabul, but it was difficult to win a lasting success, not least because of the fractured nature of Afghan politics, which made it hard to find an effective central authority with whom to make peace.
RIGHT

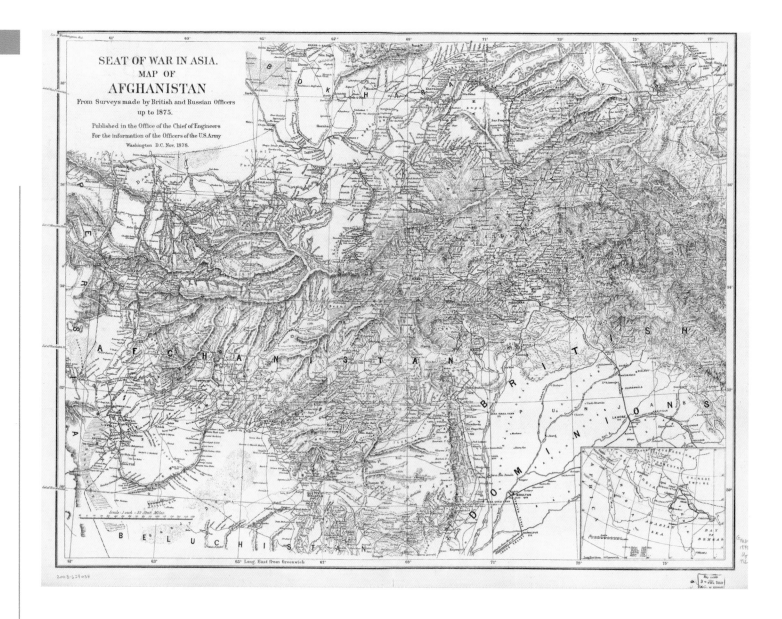

century, of still more far-flung systems. That by Belgium around Liège was to delay the German invasion in 1914. France had similar systems around Belfort and Verdun. In turn, these defensive systems were challenged by the development of artillery, notably pneumatic recoil mechanisms, rifled steel breech loaders, and delayed action fuses, which entailed new commitments in the shape of rail links capable of moving very heavy artillery pieces.

Again, there were implications for mapping and, in particular, a need in preparing for combat and during conflict for specialised military mapping that was different in character to its civilian counterpart. Firearms capability was the crucial driver of change. The relationship with topography was now over a

larger area, and this relationship encouraged an overlap; in conflict and planning, between tactical and operational levels and criteria, and a marked contrast with strategic concerns. The development of warfare in the nineteenth century led to a sense that change was normal and, as it equated with development, necessary. Military mapping and the use by the military of maps (which was not the same thing) were affected by this process and aspects of it. Maps appeared to be a way to clarify, indeed unlock, tactical, operational and strategic issues. Their use also provided a key element of training; one that could then be applied for further aspects of training. However, the demands on map use and map-makers that were to be placed by the First World War were not yet anticipated.

Zulu's first seen in great force

LINES OF ZULU ATTACK at 10 A.M.

Line of Zulu flank taken by Lord Chelmsford in the early morning of the

Road to Nthlazakazi Mtn & Malakata Hill. 10...

Kraal

High Kopjie

Russell's Rocket Batty. supported by Colᵗ Durnford FIRST ATTACK

Wide Spruit

Kraal

SECOND ATTACK of Colᵗ Durnford

2 Troops of N. Cavalry N. Carbineers, Police & Volunteers

Line of Advance of Zulu Centre

Line of Zulu flank Attack

HIGH HILLS

Great Open Valley

British line of Skirmishers

Grassy Open Country

Rough & Stony Ground

Deep Donga or Water cutting. defended by the 24ᵈ Regᵗ during the morning

BRITISH LAST POSITION Janᵞ 22ⁿᵈ

3 Troops of N. Cavalry supported by 4 Compˢ of N. Contingent (Captᵗ Barton.) retired about 9 A.M.

Native Contᵗ

24ᵗʰ

MOUNTED TROOPS

Comᵗ Lonsdale

24ᵗʰ & 6ᵗʰ

Major Durnford 22ⁿᵈ 21ˢᵗ Janᵞ

BRITISH CAMP

Kraal

first position of the 2 Guns left in Camp

2-3ʳᵈ

Nᵒ N.Cᵗ

1-3ʳᵈ

Nᵒ N.Cᵗ

2-24ᵗʰ

6 Guns 5ᵗʰ Brigᵈᵉ N. Batt. R.A

Janᵞ 21ᵗʰ & 22ⁿᵈ

MOUNTED POLICE

MOUNTED INFANTRY

NATAL CARBINEERS

1-24ᵗʰ

Col. Pulleine

Line of f...

Precipi...

Line of after the Zu... cut off the Ro... Rorke's Drift

Waggons

HEAD Qᵗˢ

HOSPITAL TENTS

Waggons

Mᵗ ISANDULA Steep & Rocky

Line of Zulu rear Attack

Slightly cultivated

Very rough ground

Kraal

Kraal

Line of Advance of Col. Durnford's mounted reinforcements 22ⁿᵈ & 23ⁿᵈ Janᵞ

Waggon Road from Rorke's Drift 12 Miles

Stony

Kraal

Ground

Running Spruit

NEW PUBLICATIONS
Wyld's large scale Military Map of Zulu Land in sheet 4/- case 5/-
South Africa .... 5/- ... 7/6
British South Africa ... 3/- ... 4/6

SKETCH OF THE ATTACK ON THE CAMP OF ISANDULA January 22ⁿᵈ 1879.

BRITISH POSITIONS ▭
ZULU ...... DO ...... ▦▦▦

ATTACK ON THE CAMP OF ISANDLWANA, BY JAMES WYLD, 1879 A map of a British disaster on 22 January 1879 that reveals the range of the Zulu attacks, although it could not capture their speed. A 20,000-strong Zulu force defeated a British force of 1,800, of whom only 581 were regulars. The Zulus enveloped the British flanks and benefited from their opponents running out of ammunition, but, thanks to the British Martini-Henry rifles, Zulu casualties were very high. The Zulus, who did not want rifles, referred to the British as cowards because they would not fight hand to hand. The map was published by James Wyld, Geographer to Queen Victoria. As an instance of the entrepreneurial world of map publishing, it included an advertisement for his maps of South Africa and the Zulu War. LEFT

BATTLE OF TEL-EL-KEBIR, 13 SEPTEMBER, 1882.

SIGNED BY MAJOR H. ARDAGH AND LIEUTENANT A.

B. CRABBE   This printed hand-drawn map
shows the Battle of Tel-el-Kebir in Egypt
in 1882. Under the command of
Major-General Sir Garnet Wolseley, the
model for Gilbert and Sullivan's 'Modern
Major-General', the British launched a
successful dawn attack on the Egyptian
position on 13 September 1882,
effectively ending the war. The British
attacked the Egyptian earthworks
without any preliminary bombardment.
Wolseley preferred to try to gain the
advantage of surprise, and his infantry
attacked using their bayonets. The
Remington rifles, Gatling machine guns
and Krupp artillery of the Egyptian army
did not help them. **RIGHT**

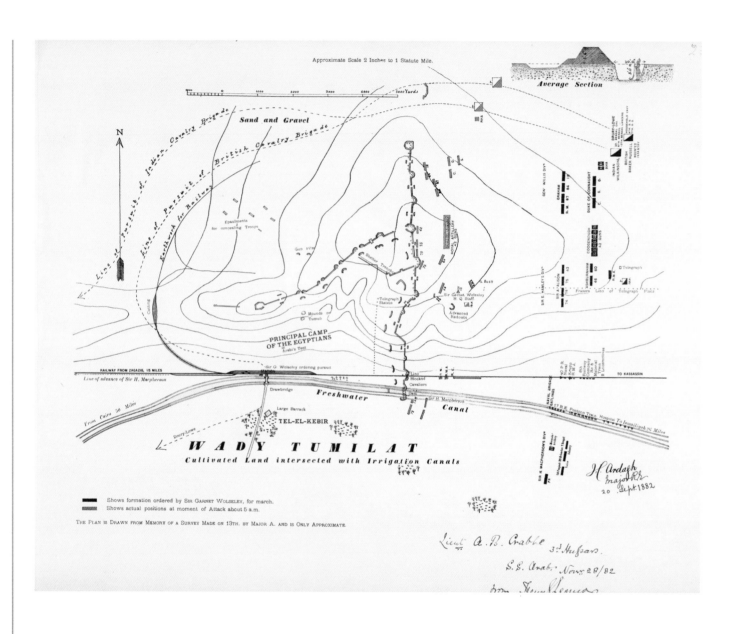

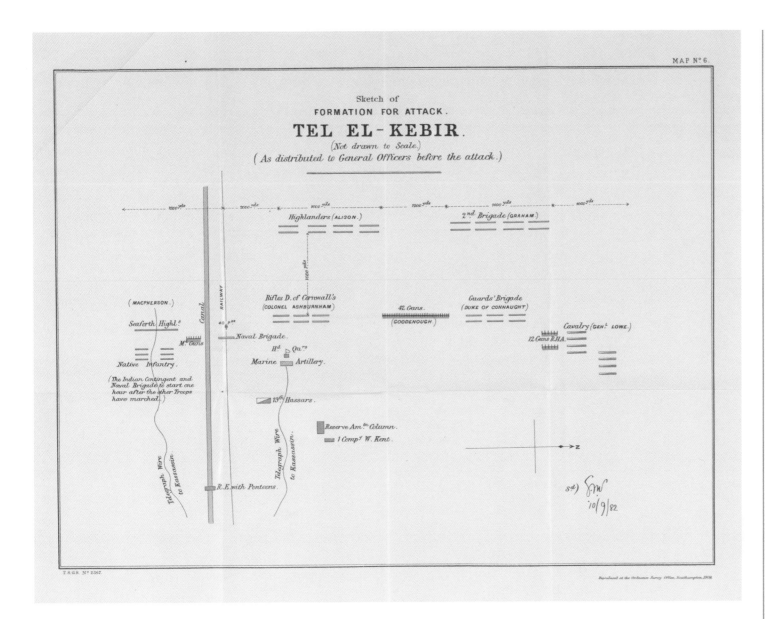

MAP Nº 6.

Sketch of
FORMATION FOR ATTACK.

TEL EL-KEBIR.

(Not drawn to Scale.)
(As distributed to General Officers before the attack.)

T.S.G.S. Nº 2367.

SKETCH OF FORMATION FOR BRITISH ATTACK, TEL EL-KEBIR, EGYPT, 1882, AS SUBSEQUENTLY PRINTED

The British attacked the Egyptian earthworks without any preliminary bombardment. Sir Garnet Wolseley preferred to try to gain the advantage of surprise and his infantry attacked using their bayonets. **LEFT**

BATTLE OF SON TAY, 1885. CHINESE SCHOOL, NINETEENTH CENTURY This war between France and China in 1883–85 that arose as a result of French expansion into Indo-China was a victory for the French and was followed by their annexation of Tonkin. French amphibious capability and firepower were important elements, but so was the marginal nature of the struggle for China and the degree of exhaustion arising from civil war and rebellions in the 1850s to 1870s. ABOVE

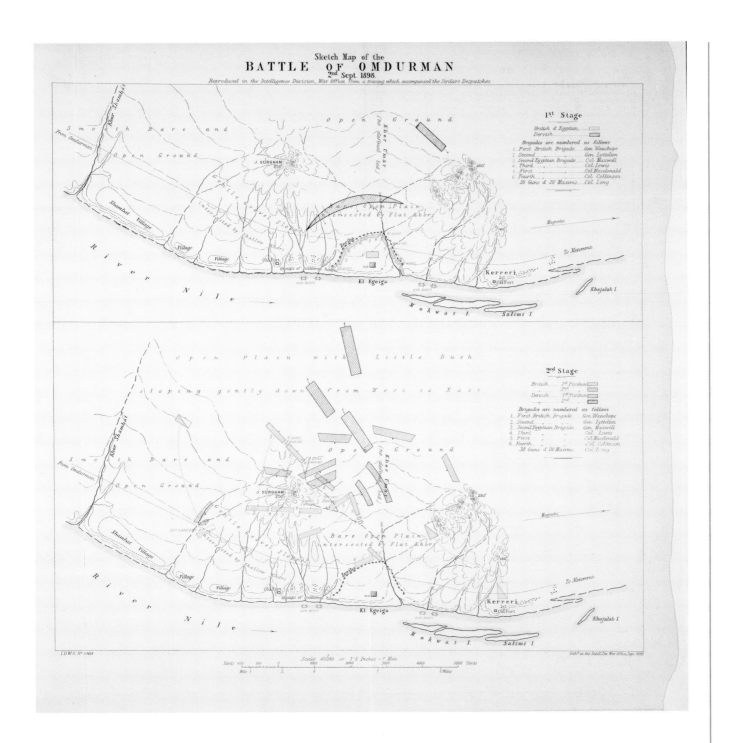

Sketch Map of the
BATTLE OF OMDURMAN
2nd Sept 1898
Reproduced in the Intelligence Division, War Office, from a tracing which accompanied the Sirdar's Despatches

BATTLE OF OMDURMAN, BY MAJOR TALBOT, 1898

Two maps on one sheet, showing the British, allied Egyptian and hostile Dervish (Mahdist) forces at the first and second stages of the battle on 2 September 1898 and the nature of the terrain. The maps were surveyed by Major Talbot of the Royal Engineers six days after the battle and lithographed at the Intelligence Division of the War Office later that month. Churchill, who was present, commented: 'It was a matter of machinery.' Artillery, machine guns and rifles devastated the attacking Mahdists. Kitchener's army had eighty cannon and forty-four machine guns, and the infantry was effective at 1,500 yards. In the first stage, the defenders used firepower from behind a *zariba* (thorn-wall) before advancing in the second stage, exposing themselves to the Mahdists reserves and being put under pressure by Mahdist attacks. Of the 26,000 troops in Kitchener's army forty-eight were killed and 434 wounded; of about 52,000 in the army of the Khalifa Abdallahi about 9,700 were killed, 10,000–16,000 wounded and 5,000 captured. The Mahdists' inappropriate strategy and tactics, not least the decision to fight a pitched battle and the frontal advance across flat ground, contributed greatly to British victory. **LEFT**

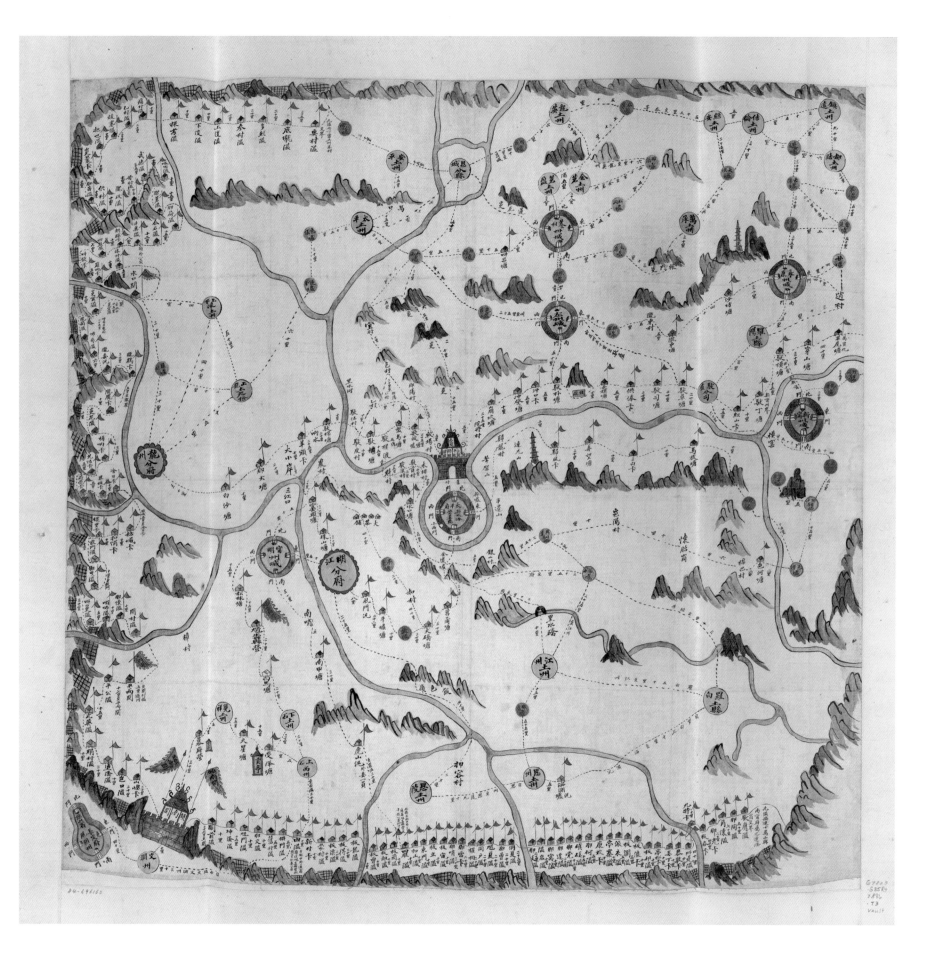

MAPA ILUSTRADO DE LA ISLA DE CUBA

Publicado por la Casa M. MAUCCI
BARCELONA—Conde Asalto, 8
1896

(Reservados los derechos de propiedad.)

GUANGXI PROVINCE, 1886 This map of part of Guangxi province shows garrison troops stationed along the pass of Zhennan Guan in Guangxi near the border with Vietnam in 1886. It reflected long-established patterns in Chinese cartography, whereas its Western counterpart was changing rapidly, notably with the development of contouring. LEFT

CUBAN INSURGENCY, 1896 An insurgency against Spanish rule in 1896, the second major one in Cuba that century, highlighted the problems of both guerrilla and counter-insurgency warfare. The insurgents hit the economy, destroying the sugar and tobacco crops, in an attempt to create mass unemployment and to force refugees to become a burden on the Spaniards. The insurgents were initially successful in evading attempts to engage them in battle, in part because the Spanish forces attempted to cover the entire country. This was a war in which raids and ambushes played a major role and in which control over supplies was important. In turn, the Spaniards sought to fix their opponents, so that they could use their firepower. Under Valeriano Weyler, they combined defensive lines, effective field operations and the movement of people to the towns. The insurgents were driven from west Cuba in 1896. Successful American intervention in 1898, which led to the end of Spanish rule, prevents us from knowing how long Weyler's policies would have achieved results. ABOVE

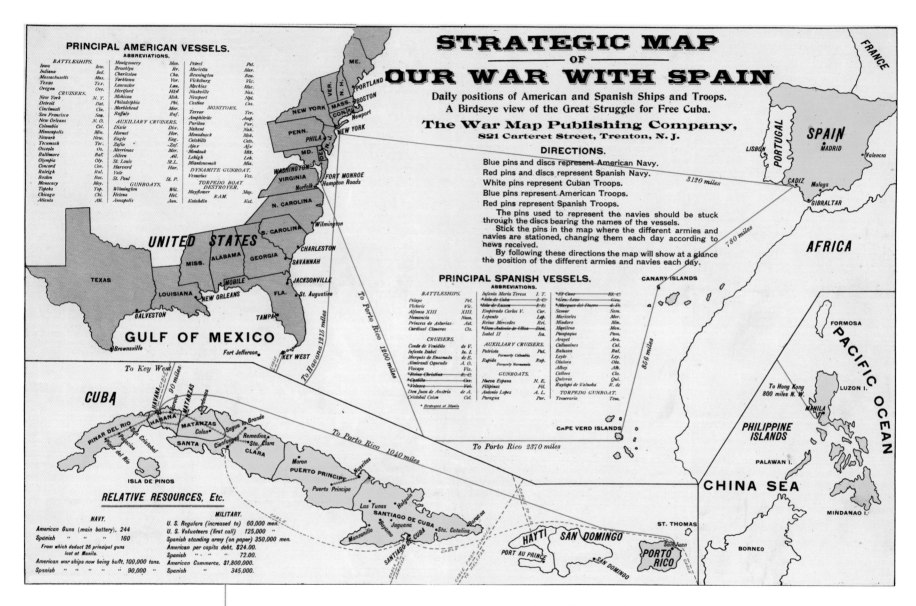

MILITARY SKETCH MAP
TO SHOW
SIR R. BULLER'S ADVANCE
FROM CHIEVELEY
TO RELIEVE LADYSMITH
FEBY 14TH TO MARCH 1ST 1900.

SIR REDVERS BULLER'S ADVANCE TO RELIEVE LADYSMITH, MILITARY SKETCH MAP, 1900 A printed work reflecting public interest, this sketch map by H. Delmé-Radcliffe is supported by illustrations of the landscape. The Boer sieges of Kimberley, Ladysmith and Mafeking led to a loss of Boer momentum that threw away the initiative gained by beginning the conflict with the British. After initial failures at Colenso, Spion Kop and Vaal Krantz, Buller relieved Ladysmith. LEFT

SEAT OF WAR IN AFRICA, BY THE AMERICAN WAR DEPARTMENT, 1899 This map of the Boer War was produced by the military information division in the Adjutant General's office of the American War Department. It testified to the growing international interest in wars involving other states, in part in order to try to gain knowledge by such means. Military observers became more common. American interest in part reflected the new engagement with out-of-area conflicts seen the previous year with the war with Spain. Provision of mapping for the army was put on a more systematic footing in 1909 when a map reproduction unit was established at the Army War College. RIGHT

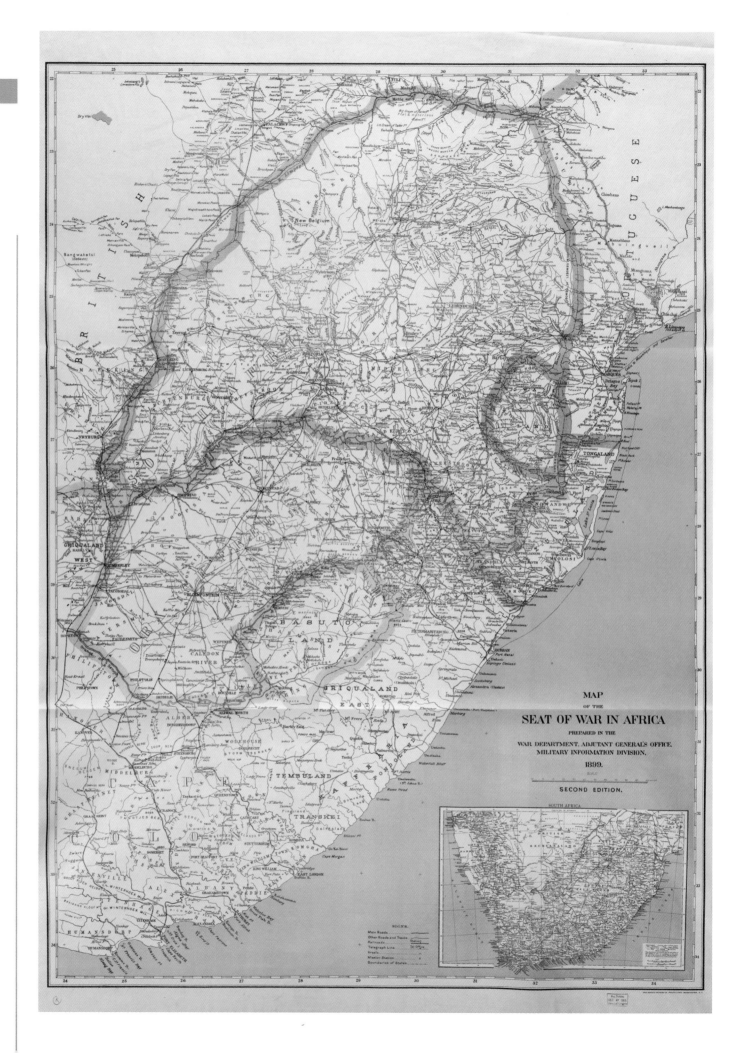

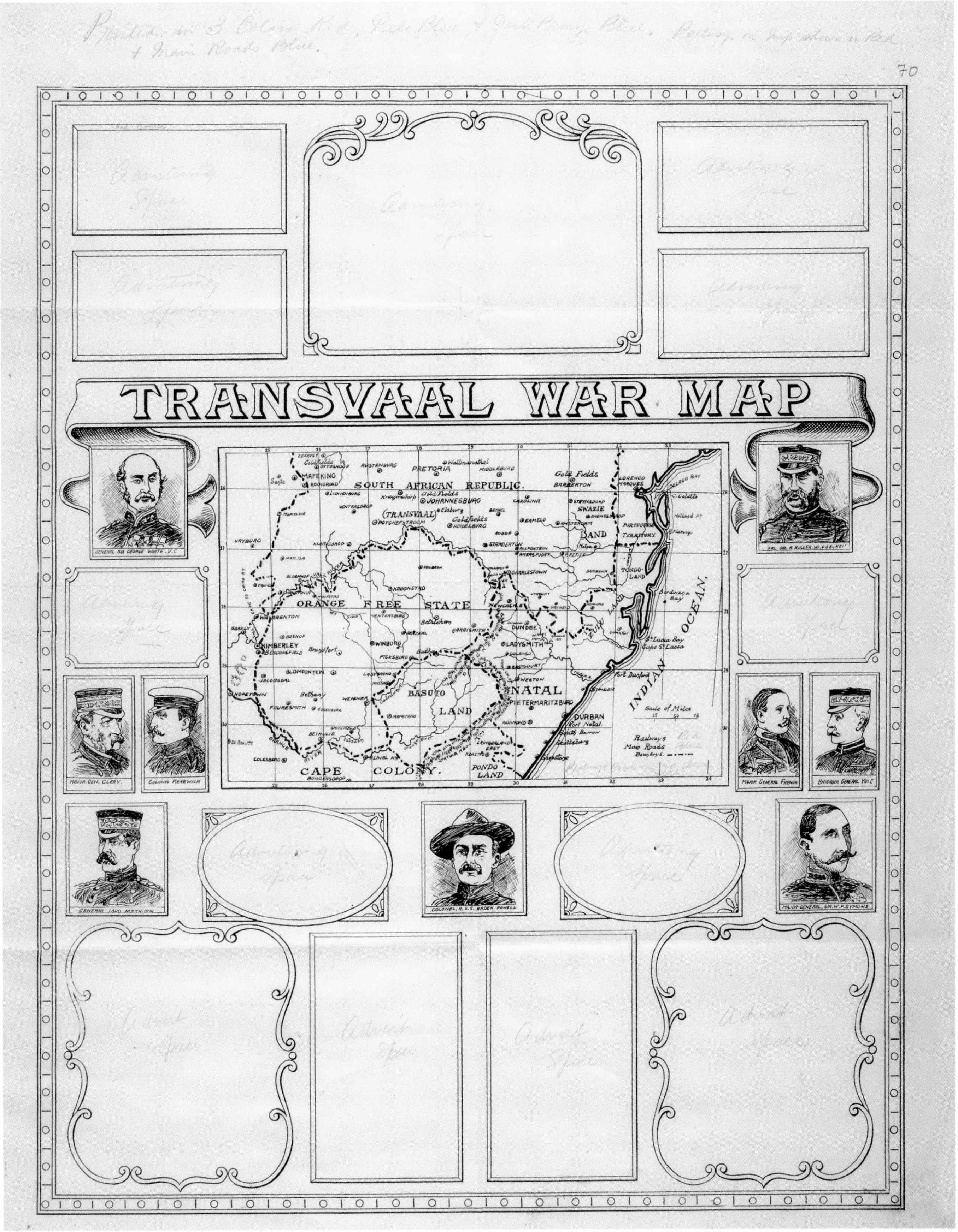

TRANSVAAL WAR MAP, BY THE STATIONERS'
COMPANY, 1899 Created by the Copyright
Office of the Stationers' Company, this
was not, however, a useful map as there
was no guide to the terrain. The space
left for advertisements indicated the
commercial use of this map. Illustrations
of commanders were an easy way to fill
space. LEFT

SIEGE OF KIMBERLEY, BY COLONEL A.J. O'MEARA,

1900 This map from the 1899–1900 Boer War, with pictorial compass points showing speared Boers, was surveyed by Colonel A.J. O'Meara, of the Royal Engineers, signed by Claude Lucas, Cape Government Surveyor, in February 1900, and published in South Africa that year. By taking the initiative, the Boers put much pressure on the British that winter. Serious deficiencies in British tactics and training included a continued preference for frontal attacks and volley firing, a lack of emphasis on the use of cover and a lack of appreciation of the enhancement of defensive firepower, not least due to the use of smokeless powder. The British suffered from a lack of adequate maps and were generally forced to rely on poor-quality compilations produced by fitting together the title diagrams of inaccurate field surveys. This encouraged the formation of a topographical section for the General Staff established in 1904. The Boers also lacked maps. RIGHT

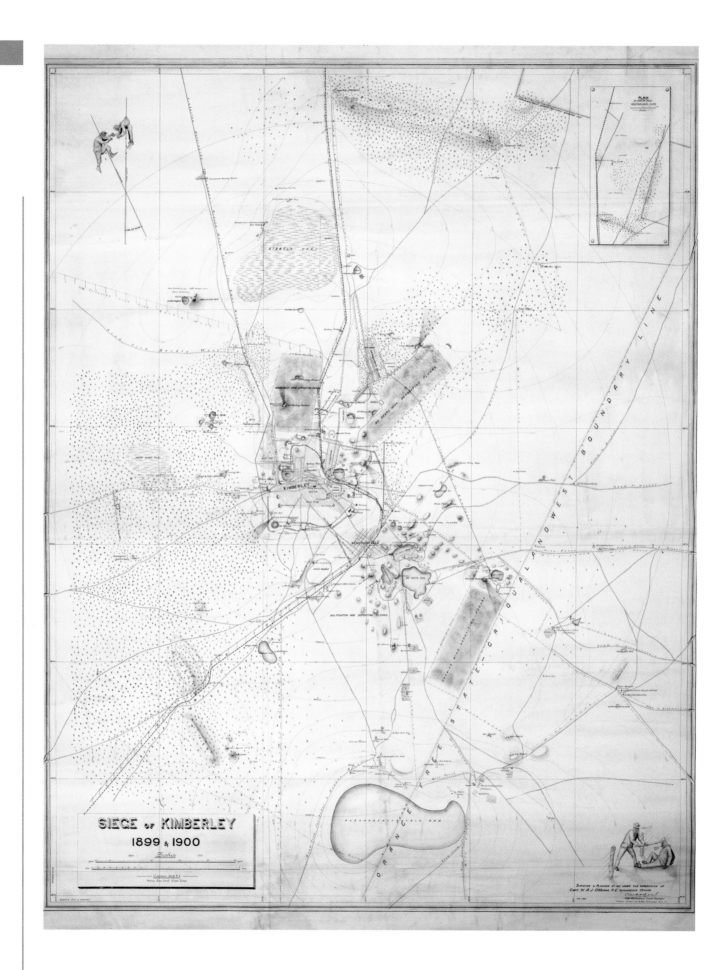

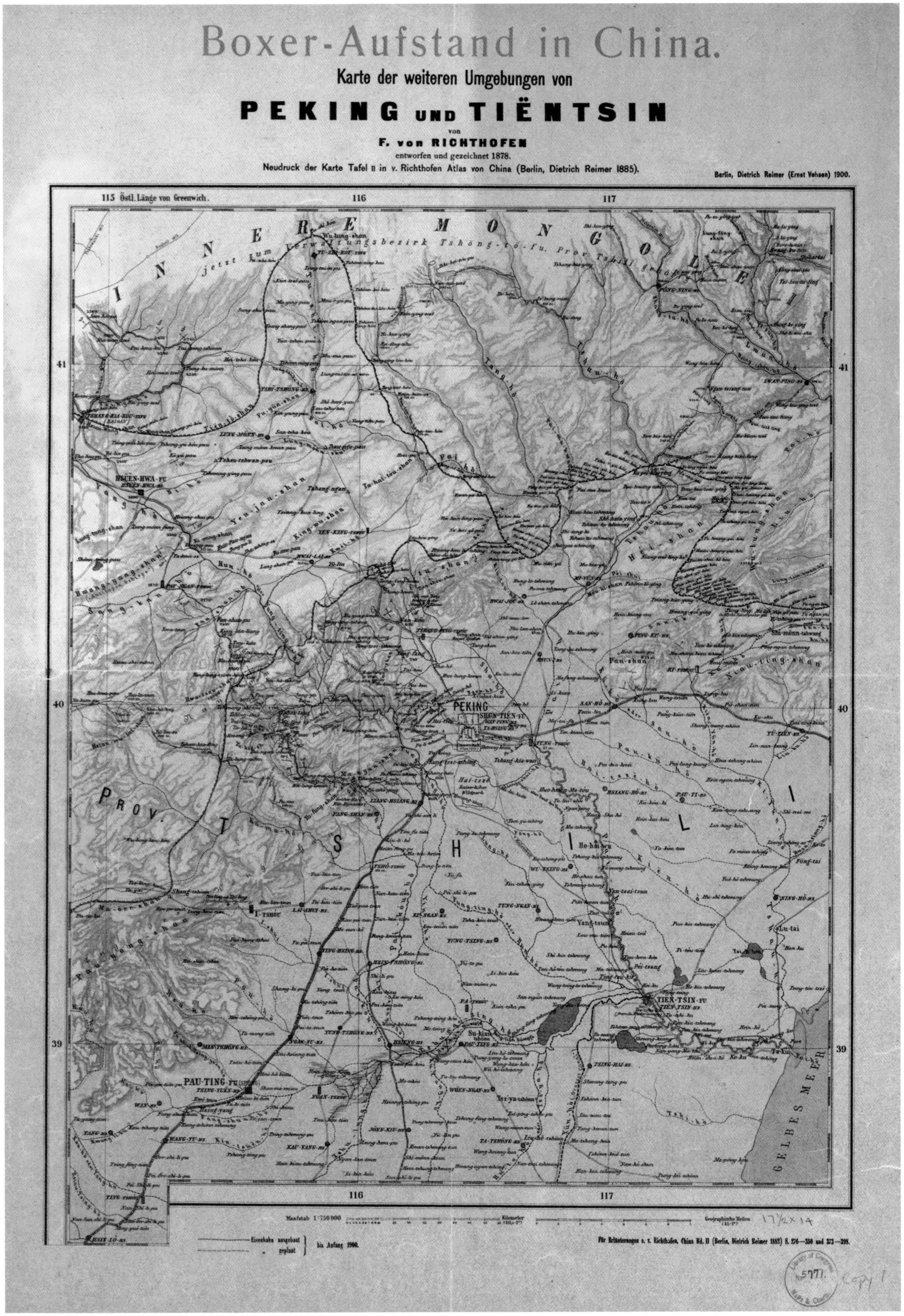
Boxer-Aufstand in China.
Karte der weiteren Umgebungen von
PEKING UND TIËNTSIN
von
F. von RICHTHOFEN
entworfen und gezeichnet 1878.
Neudruck der Karte Tafel II in v. Richthofen Atlas von China (Berlin, Dietrich Reimer 1885).
Berlin, Dietrich Reimer (Ernst Vohsen) 1900.
115 Östl. Länge von Greenwich.
116
117
41
40
39
INNERE MONGOLEI
PEKING
SHUN-TIEN-FU
PROV. TSHILI
TIEN-TSIN-FU
PAU-TING-FU
GELBES MEER

THEATRE OF OPERATIONS IN CHINA, BY FERDINAND VON RICHTHOFEN, 1900 This map was published in Germany in 1900 and subsequently acquired by the Library of Congress after the Second World War. The Boxer movement became nationally and internationally significant in 1900 when the imperial court increasingly aligned with the Boxers against foreign influence. Converts to Christianity were killed, hostilities between Boxers and foreign troops began, the government declared war on the foreign powers, and the foreign legations in Beijing (Pekin on map) were attacked on 20 June and then besieged. In response, international relief expeditions were mounted. The first was blocked by the Chinese and forced to retreat to Tianjin (Tientsin), which was unsuccessfully besieged by the Boxers, whose swords and lances, however, provided no protections against the firearms of the Western garrison. After Tianjin had been relieved, a second force was sent to Beijing, defeating opposition en route, breaching the walls of the city and relieving the legations on 14 August. The alliance of Western and Japanese troops paraded through the Forbidden City, a powerful sign of Chinese loss of face. The subsequent treaty with China in 1901 decreed very large reparations, as well as twelve foreign garrisons between the coast and Beijing. LEFT

MAP OF CHINA, FROM *LE PETIT JOURNAL*, 1900

An illustration from *Le Petit Journal* on 8 July 1900, showing the prime theatre of operations for the Boxer Rising. The significance of communication routes is shown with the mapping of telegraph cables and railways, existing, in construction or envisaged. The Boxer Rising, a violent anti-foreigner movement that began in 1897, was countered by an eight-nation international force. RIGHT

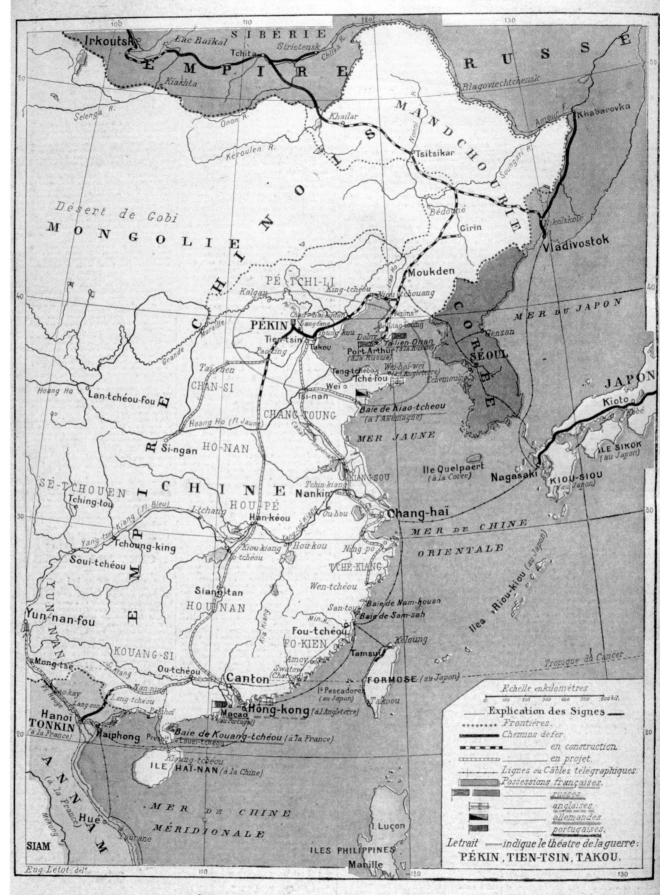

LES ÉVÉNEMENTS DE CHINE
Carte du théâtre de la guerre

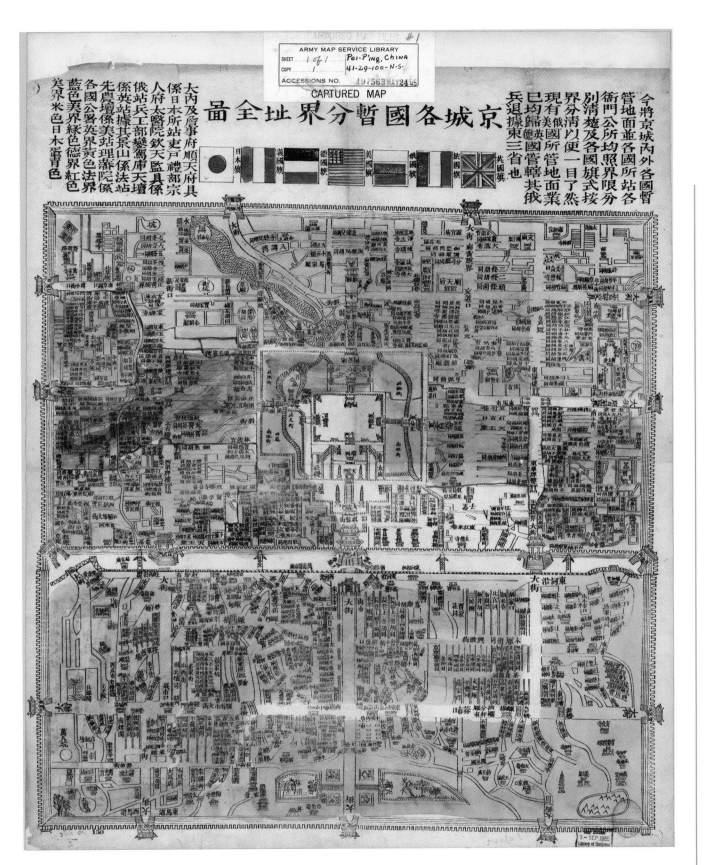

FOREIGN LEGATIONS IN BEIJING DURING THE BOXER REBELLION. PUBLISHED IN BEIJING, C. 1900 Britain, France, Germany, Japan and the USA are depicted in different colours. The Boxer siege of the legations was unsuccessful, but they were put under pressure, not least because the Chinese had a Krupp quick-firing cannon. **LEFT**

TOTAL WAR

FIRST BATTLE OF THE MASURIAN LAKES 5–15 SEPTEMBER 1914 East Prussian Landwehr (reserve forces) of the 17th Army Corps lie low with their rifles in a trench near Lötzen. **PREVIOUS PAGES**

ACROSS THE TWENTIETH CENTURY, wars greatly increased the military use of maps, as well as public interest in them. The two processes were linked, but also separate. In the first place, combatants used and produced large numbers of maps, numbers far greater than those produced for earlier conflicts. In the twentieth century, this process began prior to 1914, because the wars of that period led to much mapping. For example, in confronting the serious Boxer Rebellion in China in 1900, the British drafted maps there, printed them in India and then sent them to China for use, benefiting greatly from the speed provided by steamships and their capacity to integrate empires.

The large-scale Russo-Japanese War of 1904–05 saw the Russians pushed back from the area of southern Manchuria they had already mapped. As a result, under Japanese pressure, they turned to map northern Manchuria. In the Balkans, the wars of 1912–13, which resulted in the Turks losing much of their European empire, led to mapping, as did the Italian invasion of Libya in 1911.

GEOPOLITICS

The use of maps also proved significant for the new concept of geopolitics. Although the idea preceded the word, the term geopolitics was devised at the end of the nineteenth century and then used thereafter in order to provide apparent scientific credibility for ideas of shaping a synergy of power and territory. These ideas were to be advanced in particular by German commentators in the 1920s and 1930s. However, already, the use of geopolitical concepts was important in the 1900s and 1910s, as writers and others sought to discern opportunities and threats in the international arena.

The most prominent individual instance, Halford

Mackinder's 1904 lecture and essay on the changing geopolitics of Eurasia, did not use the specific word, but it very much linked strategic possibilities to changing technology. Indeed, the immediate reception of the lecture saw a tension between the possibilities of rail and those of air. Mackinder used maps to illustrate his arguments, both in this piece and in other publications. While not directly a case of mapping for war, the use of maps by geopoliticians encouraged the idea that maps were part of the intellectual armoury that had to be deployed when international power politics was considered.

The unprecedented range, scale and intensity of the First World War (1914–18) ensured that it proved particularly significant in the development of mapping. In part, this significance was a reflection of the deficiencies with existing maps. This deficiency proved especially acute outside Europe, where the mapping for conflict was generally poor, and notably so if there had been no large-scale conflict already.

Unluckily for the British from this perspective, South Africa, where they had fought the Boer War in 1899–1902 leading to extensive (and late) post-war mapping, was not again an area of conflict. Instead, the British and French found themselves attacking the German colonies in Africa for which they had no good maps. In German East Africa (later Tanganyika, and then most of Tanzania), the British navy knew in 1914 from signal intercepts that the German cruiser SMS *Königsberg* was at Salale, but Salale was not marked on the naval charts, and it took eleven days to identify it. Such a situation encouraged conflict near landmarks.

Moreover, a lack of adequate maps of Africa made it difficult to predict terrain or watercourses. On the other hand, the deficiencies revealed during the Boer War had led to a post-war improvement in British

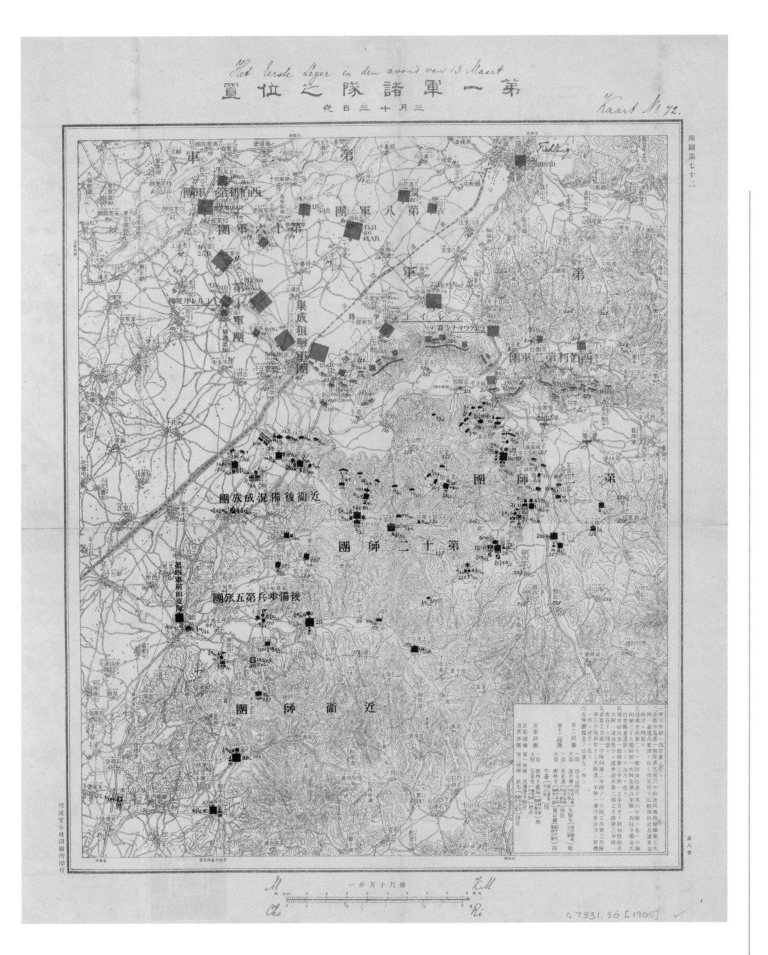

Het Eerste Leger in den avond van 13 Maart

Kaart Nº 72.

G 7831.56 [1905]

RUSSO-JAPANESE WAR, 1904–1905 Japanese map of the area around Tieling, Kianoning, China, on 13 March 1905, after the battle of Mukden (19 February–10 March 1905). Japanese forces shown in blue and Russian forces in red. It was the last land battle of the Russo-Japanese War, with each side deploying about 300,000 troops along a nearly 50-mile front. The Russians beat off Japanese attacks before retiring as they feared being outflanked and encircled. The Japanese frontal assaults on entrenched forces strengthened by machine guns and quick-firing artillery prevailed despite horrific casualties. Success for the side that took the initiative was the lesson taken into the First World War. LEFT

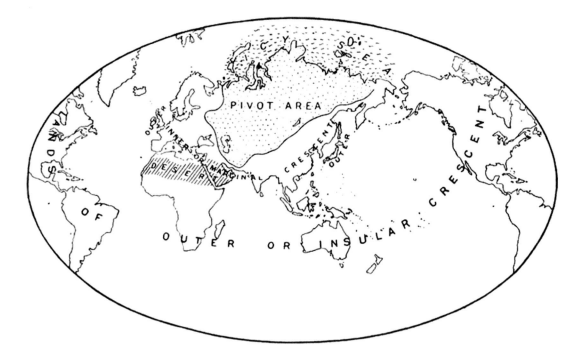

'THE NATURAL SEATS OF POWER', BY HALFORD MACKINDER, 1904 In Halford Mackinder's 1904 account of geopolitics he developed his notion of a Eurasian heartland that was impregnable to attack by sea but a threat to the whole of Eurasia. He claimed that in the heart of Eurasia there was a 'geographical pivot of history'. Railways, he argued, had made the heartland power, now Russia, more potent. They were presented as key demonstrations and enablers of power. Mackinder pressed for a united British empire able to resist the heartland power. He argued that 'it is desirable to shift our geographical view-point from Europe, so that we may consider the Old World in its entirety' and that 'in the world at large she [Russia] occupies the central strategical position held by Germany in Europe'. In practice, Britain, France, Germany and Russia overestimated Russia's military potential, as the First World War was to show.

ABOVE

military surveying. This ensured that British military surveying during the First World War was much more impressive than it would otherwise have been.

Problems with maps also faced Turkish operations in the Caucasus in 1914 and German planning for a Turkish invasion of Egypt in 1915. Mesopotamia (Iraq) and Palestine were also very poorly mapped. Indeed, only the European portion of the former Turkish empire was mapped reasonably well, or comparatively so. This encouraged irregular units in the Asian portion, notably in the Arab rising against the Turks in the Hijaz (western Arabia) to trust their 'mental' maps.

Secondly, there were the problems of interpreting existing maps of Europe. This was crucially seen with the German invasion of Belgium and France in 1914. Maps had to be read in light of how swiftly the opposing forces were able to move, which only partly depended on the terrain or topography, and on how difficult they would find it to mount an offensive. The French, for example, could, as they did in 1914, readily use their rail system to move troops from eastern France, where they had been unsuccessfully attacking German troops, to support the defence of Paris. The Germans, on the other hand, faced the problems of taking over the rail systems of conquered areas of

Belgium and France, including different gauges of track. The use of maps did not necessarily cover such issues (which, instead, led to specialised diagrams) and, moreover, did not always ensure that commanders could locate their units as accurately as might be anticipated.

TRENCH WARFARE

The particular needs of trench warfare created new demands for mapping. These needs had not been anticipated by military cartographers. The French, for example, had concentrated their military mapping on fortified positions, such as Belfort, only to discover that most of the fighting took place nowhere near these. They were not prepared for the mobile warfare that occurred within France in 1914. Instead, it became necessary to respond first to this warfare and then, from late 1914, to the need for detailed trench maps, in order to be able to plan both effective defences and successful assaults on them. The hard-fought First Battle of Ypres, in October and November 1914, a bitter struggle waged in a relatively small area in which the Germans failed to break through, led the British to appreciate this need.

Effective infantry–artillery coordination was important to success in attack and defence. In trench warfare, accurate surveying and mapping reduced the need for the registration of targets by guns prior to attack and therefore allowed an element of surprise, which was important to the success of offensives.

Moreover, maps were crucial in order to produce coherence in command, in particular the implementation of strategic plans in terms of timed operational decisions and interrelated tactical actions. Without an adequate grasp of maps and what they meant, command practices were unsystematic and less than taut. Such features encouraged incoherent

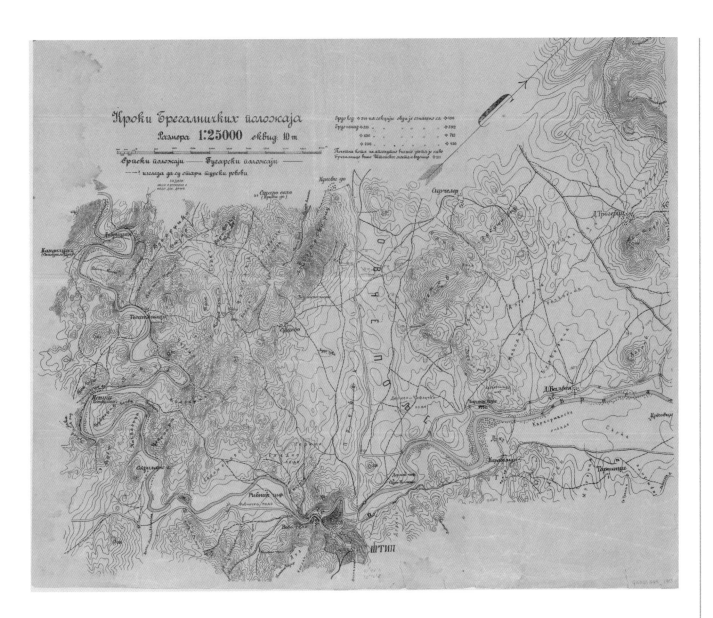

strategies, battles without effective overall plans and piecemeal tactics. It became necessary to excel in the midst of uncertainty.

When the British Expeditionary Force (BEF) was sent to France in 1914, one officer and one clerk were responsible for mapping, and the maps were unreliable. The three existing survey sections remained with the Ordnance Survey in Britain. By 1918, the survey

organisation of the now much larger BEF had risen to about 5,000 men and had been responsible for more than 35 million map sheets. No fewer than 400,000 impressions were produced in just ten days in August 1918.

Due to the nature of trench warfare, maps were produced for the military at a far larger scale than those with which they had been equipped for mobile campaigning. The British army replaced its 1:80,000 scale maps with 1:10,000 scale maps, a formidable task of both map-making (in the midst of war) and production. These maps needed to be up to date and required a high degree of accuracy in order to permit indirect fire, as opposed to artillery firing over open sights. This accuracy came in part from better guns but mostly from improvements in photogrammetry (mapping by means of photography), which reflected the importance of aerial information.

USE OF AIRCRAFT

Indeed, the breakthrough in surveillance came from manned flights, which began in aircraft (as opposed to balloons) in 1903. Cameras, mounted first on balloons, then on aircraft, were able to record details and to scrutinise the landscape from different heights and angles. Instruments for mechanically plotting from aerial photography were developed in 1908, while a flight over part of Italy by Wilbur Wright (one of the two Wright brothers) in 1909 appears to have been the first on which photographs were taken. The range, speed and manoeuvrability of aircraft gave them a great advantage over balloons, and they were far less vulnerable to defensive fire or to attack by other aircraft.

With the First World War came air–artillery coordination. At first, maps did not play much of a role in this. Thus, a report in *The Times* on 27 December 1914 noted of the Western Front in France and Belgium:

The chief use of aeroplanes is to direct the fire of the artillery. Sometimes they 'circle and dive' just over the position of the place which they want shelled. The observers with the artillery then inform the battery commanders – and a few seconds later shells come hurtling

on to, or jolly near to, the spot indicated. They also observe for the gunners and signal back to them to tell where their shots are going to, whether over or short, or to right or left.

However, with time came more static positions, concealed locations to protect the guns and a need for heavier and more precise artillery fire. These led to the use of maps as the key means of precision and planning. For the Battle of Neuve Chapelle on 15 March 1915, the first British trench warfare attack, the plan was based on maps created through aerial photographs.

The invention of cameras able to take photographs with constant overlap, proved a technique that was very important for aerial reconnaissance and thus surveying, notably with the development of three-dimensional photographic interpretation. Maps worked to record positions, as well as to permit the dissemination of the information. The ability to build up accurate models of opposing trench lines was but part of the equation, one made more significant by the need for aerial surveillance in order to understand the depth of defence provided by multiple trench lines. Thus, in September 1915, in the Battle of Loos, the British broke through the first German position but were stopped by the second one. The greater use of artillery that became more possible with detailed maps also changed the landscape at a tactical level with the proliferation of shell craters.

It was also necessary to locate the position of artillery in a precisely measured and understood triangulation network. This network and location permitted directionally accurate long-range artillery by means of firing on particular coordinates. Lieutenant-Colonel Percy Worrall, a British infantry officer, noted of the Western Front in April 1918, when the German Spring Offensive was being

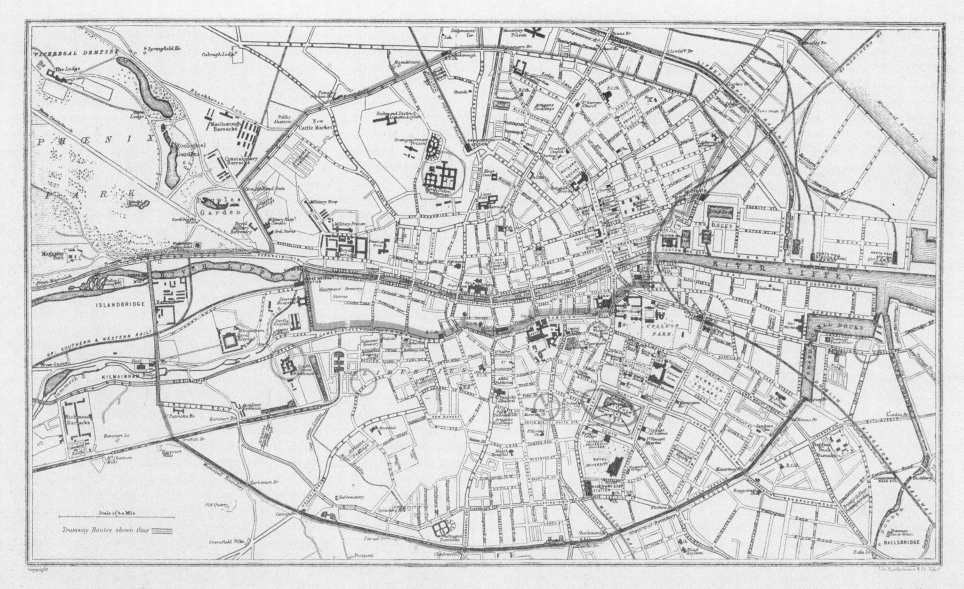

The light red line shows the cordon of troops surrounding the city; the heavy red line shows the wedge of troops which divided the forces of the rebels into two, cutting all communications. The red circles show the rebel strongholds. On the north from left to right, North Dublin Union, Four Courts and General Post Office; on the south, South Dublin Union, Marrowbone-lane Distillery, Jacob's Factory, Stephen's Green, and Boland's Mill.

DUBLIN DURING THE EASTER RISING, 1916 About 1,200 men rose on 24 April 1916, seized a number of sites and proclaimed an independent Irish Republic. However, the rebels suffered from bad planning, poor tactics and the lack of German help, as well as from the strength of the British reaction, which included an uncompromising use of artillery to shell targets in Dublin. Outside Dublin, due to divisions in the leadership, the planned nationalist uprising failed to materialise. Under heavy pressure, the insurgents unconditionally surrendered on 29 April. The firm British response was to play a major role in Irish public memory, notably in encouraging opposition to Britain. However, given the fact that Britain (including Ireland) was at war, and indeed not doing well, the declaration of martial law, the series of trials, the execution of fifteen rebels and the internment of many were scarcely surprising. Although militarily unsuccessful, the rising showed that militancy could not be contained within established political divisions and practices. The majority of the Irish population, however, loyally supported the war effort against Germany. **ABOVE**

Labels in image:
VICEREGAL LODGE
Wellington Monument
RICHMOND BARRACKS
PHŒNIX PARK
KILMAINHAM GAOL
Barracks
PORTOBELLO BARRACKS
GRAND CANAL
Royal Barracks
Workhouse
BROADSTONE STN
Seized by Sinn Feiners
ST PATRICK'S CATHEDRAL
CHRIST CHURCH CATHEDRAL
FOUR COURTS AREA
JACOBS BISCUIT FACTORY looted for supplies
FOUR COURTS Held by Sinn Feiners
Castle
CITY HALL Held by Sinn Feiners
Houses destroyed
SACKVILLE ST AREA
Barricades & Trenches
Bank
POST OFFICE Headquarters of the Sinn Feiners
Nelson Column
Rotunda RUTLAND SQ
ST STEPHENS GREEN AREA
ST STEPHENS GREEN
GRAFTON STREET
O.T.C. check attack on bank
O'CONNELL BRIDGE
LIBERTY HALL bombarded
Parnell Monument
SHELBOURNE HOTEL
MERRION ST
TRINITY COLLEGE
Warehouses
Custom House
MERRION SQ
WESTLAND ROW STN Seized by Sinn Feiners
GT BRUNSWICK ST
Warehouses held by snipers
CITY QUAY
RIVER LIFFEY
NORTH WALL QUAY
RINGSEND QUARTER

A BIRD'S EYE OF DUBLIN, BY D. MACPHERSON, 1916

Looking westward up the River Liffey, this map
drawn by D Macpherson shows positions held by
the insurgents during the Easter Rising and was
published in *The Sphere* on 6 May 1916. ABOVE

desperately, but finally successfully, resisted: 'The artillery and machine-gun corps did excellent work in close co-operation … it was seldom longer than 2 minutes after I have "X-2 minutes intense" when one gunner responded with a crash on the right spot.'

This note reflected the significance of accurate mapping. The use of grid systems on maps was part of this process; military maps came to show north points, magnetic declination and grid deviation. These guide systems aided the precise use of the increasingly precise information they presented. Each element of precision was important.

COORDINATION

More generally, the size of the armies, the proliferation of new weapons and the extent of entrenchment all forced commanders to start thinking about the coordination in time and space of fire, manoeuvre, obstacles, reserve positions and so on, largely sight unseen and accomplishing such coordination by topographic maps, aerial photography and enhanced communications. These were essential first in the defence, but also helped provide the elementary skills and infrastructure that allowed offensive ideas to grow. The British commander on the Western Front, Field Marshal Sir Douglas Haig, was critical on 3 July 1916 of his French counterpart, Joffre, stating: 'The poor man cannot argue, nor can he easily read a map.'

Maps, watches and telephones helped the coordination of manoeuvre with artillery; much synchronisation had to take place to carry an army through an enemy's defences. Nevertheless, the size and weight of radios made them cumbersome for attacking infantry units while telephone communications between command posts and the attacking infantry was not possible. To determine the actual location of an attacking infantry unit could have been difficult for the commander of the unit on site, as well as for the command post in the rear. In contrast, relatively immobile artillery units had good communications with the command posts.

The range of demands for maps increased greatly as the war continued, not least as there was a search for advantage by the competing powers. New requirements included air navigation maps for the rapidly growing number of aircraft; these included, as one form, perspective maps. The combatants on the Western Front also used hydro-geological mapping, transport planning maps and maps of the extensive underground mining.

The geographical range of mapping reflected the wide-ranging nature of campaigning. For example, the British used aerial photography on the Palestine Front and in Mesopotamia (Iraq), with the Survey of Egypt and the Survey of India respectively responsible for the bulk printing of maps. The intensity of reconnaissance photography was such that the Germans were able to produce a new image of the entire Western Front every two weeks; thus they could rapidly produce maps that responded to changes on the ground.

WARTIME CARTOGRAPHY

More widely, there was a recruitment of geography for the cause of war as there had not been in the Napoleonic Wars. The British Royal Geographical Society (RGS) played a significant role as a cartographic agency closely linked to the intelligence services. It produced maps at the request of the government. In 1914, the RGS urgently addressed the tasks of producing an index of the place names on the large-scale maps of Belgium and France issued to the British officers sent there. A four-sheet wall map of Britain at the scale of 1:500,000 was produced in order

FIRST BATTLE OF THE MARNE. FRENCH PEN AND

INK SKETCH, 1914 This map portrays the situation on 5 September 1914 on the eve of the battle. The Germans made a crucial mistake when they mishandled their advance near Paris, providing an opportunity for an Allied, principally French, counter-attack. Kluck's First Army had departed from the original plan to advance to the west of Paris, as part of a grand envelopment of the French forces. Instead, on 30 August, he decided, with Moltke's agreement, to move south, so as to advance to the east of Paris and thus maintain contact with Bülow's Second Army on his left. This axis, however, made Kluck vulnerable to the Sixth Army the French were building up near Paris. French aircraft reported the significant change of direction of the German advance. The Sixth Army was sent on 3 September to attack Kluck's right flank on the River Ourcq. In response, on 7 September, Kluck turned to the west to meet this attack, only to create a gap between his army and that of Bülow, a gap into which the British and French forces slowly advanced on 8 September. RIGHT

to help the War Office plan home defence strategies in the event of a German invasion, which was feared in 1914. The RGS then pressed on to produce a map of Europe at the scale of 1:1 million. By the end of the war, more than 90 sheets had been prepared, covering most of Europe and the Middle East. The Ordnance Survey and many commercial businesses also played a major role.

Not all combatants developed wartime cartography at the same rate. Facing operations across a wide extent of relatively poorly mapped territory, and with its limited resources under great pressure, Russia did not see a comparable production of aerial photography or the use of the resulting information for artillery

targeting. The same was the case for Austria.

INFORMATION AND PROPAGANDA

The First World War also greatly increased public interest in maps, and newspaper readers expected news coverage to be accompanied by them. Maps were employed both to provide what were intended as objective accounts and for what was consciously provided as propaganda, although the distinction between the two was not always easy to establish. In the first case, large numbers of maps were printed in newspapers in order to locate areas of conflict. They provided a more valuable addition to text than photographs and were especially valuable for the

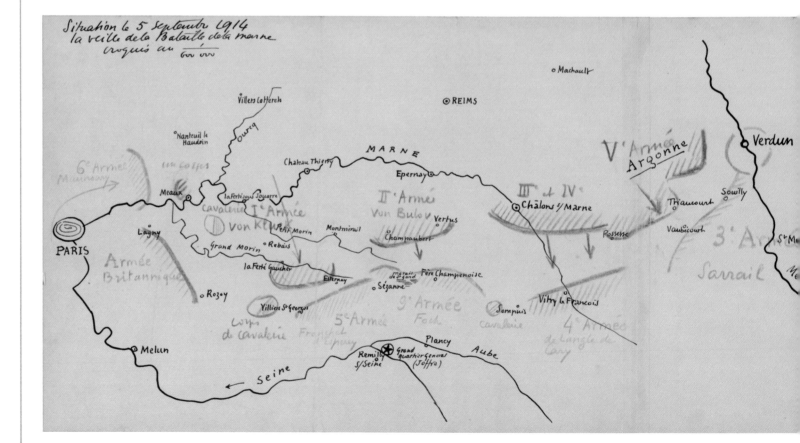

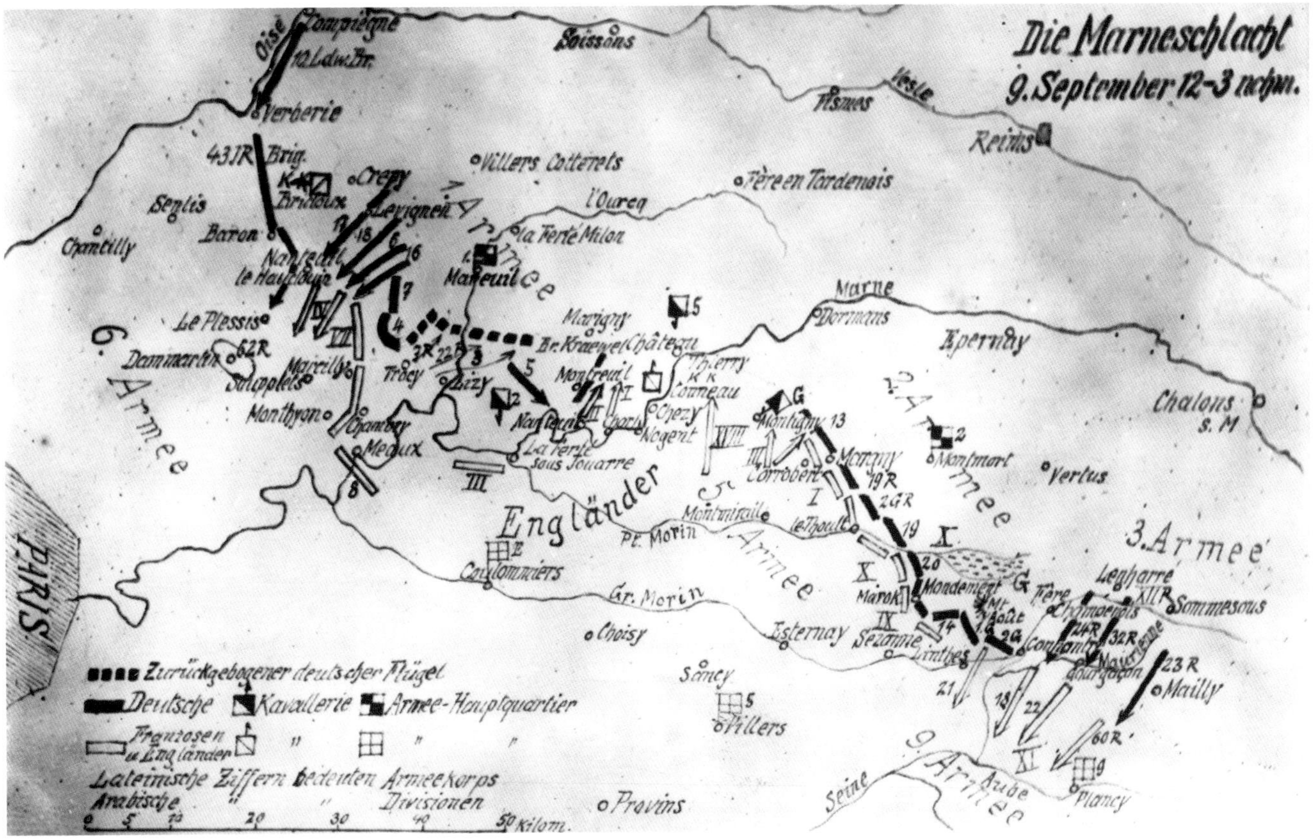

Die Marneschlacht
9. September 12–3 nchm.

THE BATTLE OF THE MARNE, FROM A GERMAN PERSPECTIVE, 1914 The advance of British and French forces on 9 September 1914 threatened both the German First and Second Armies. The Germans were not defeated and their perception, as shown in this map, was of continued advances by the First and Third Armies. However, in the face of the Allied advance, the German High Command suffered a failure of nerve and, on 10 September, Kluck was ordered to pull back. The Germans withdrew to the River Aisne in what became a key failure of impetus. They were never to be in as favourable an operational position on the Western Front. At this stage of the war, the Germans suffered from a shortage of maps and survey data. LEFT

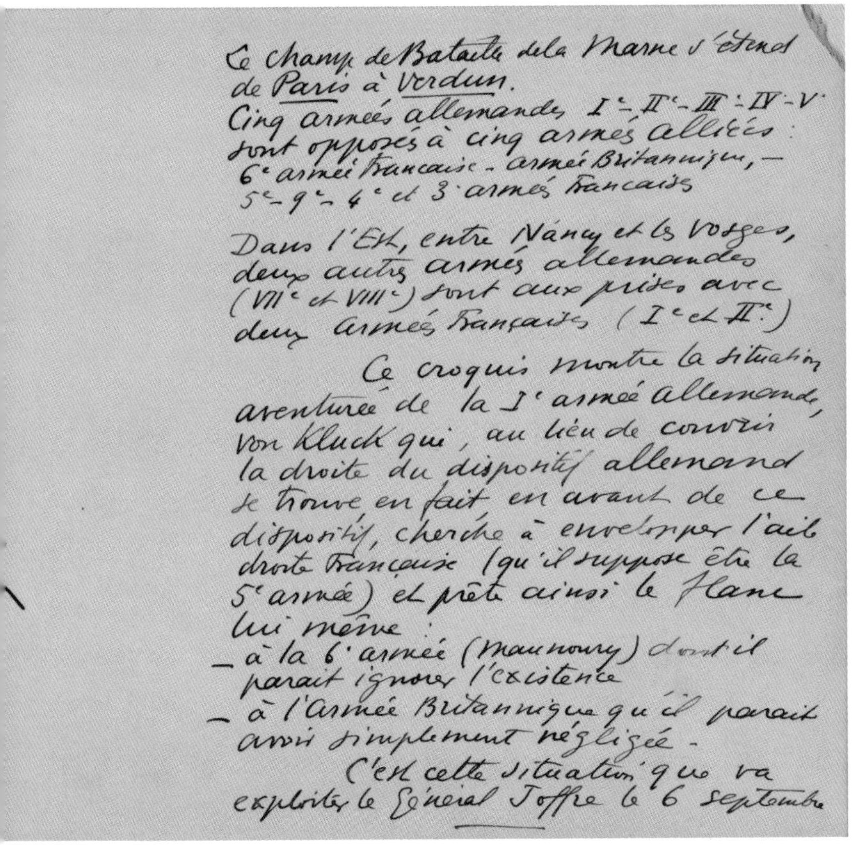

distant areas not covered in conventional atlases. They also gave the detail necessary to follow, or at least locate, trench warfare. The colour photography that was to come later in the century was not yet an established part of newspaper publishing and, as a result, black-and-white maps were not overshadowed. Providing helpful newspaper maps, however, was not easy. The simple black-and-white maps generally included little, if any, guidance to terrain, the difficulties of communications or the nature of the front line.

There was also a rapid production of atlases to satisfy strong consumer interest. These included the *Atlas of the European Conflict* (Chicago, 1914), *The Daily Telegraph Pocket Atlas of the War* (London, 1917), *Géographie de la Guerre* (Paris, 1917), *The Western Front at a Glance* (London, 1917), *Petit Atlas de la Guerre et de la Paix* (Paris, 1918) and *Brentano's Record Atlas* (New York, 1918). The theme was readily nationalistic, as with the *Album-atlas of British Victories on the Sea* (1914). Approved by the Official Press Bureau and

MAP TO ACCOMPANY BRITISH MILITARY REPORT ON SINAI PENINSULA, BY GENERAL STAFF, 1914

This map was endorsed as received by the map room of the geographical section of the General Staff. Palestine was a Turkish possession, while Egypt was under British control. The Turkish attack on the Suez Canal was repelled in February 1915 by Indian units supported by British and French warships. The Turks did not benefit from any supporting rising in Egypt. The Turkish columns advancing on the canal were spotted by British aircraft. Then the commitment to the defence of the canal was translated from defensive preparations to a forward defence in which the British moved into Sinai. However, there was no significant pressure on the Turks in Palestine until the spring of 1917 when the British focused on the Turkish positions near Gaza. The background was a major attempt to create the necessary transport and logistical infrastructure. In the autumn of 1917, the focus switched to Beersheba which was captured. RIGHT

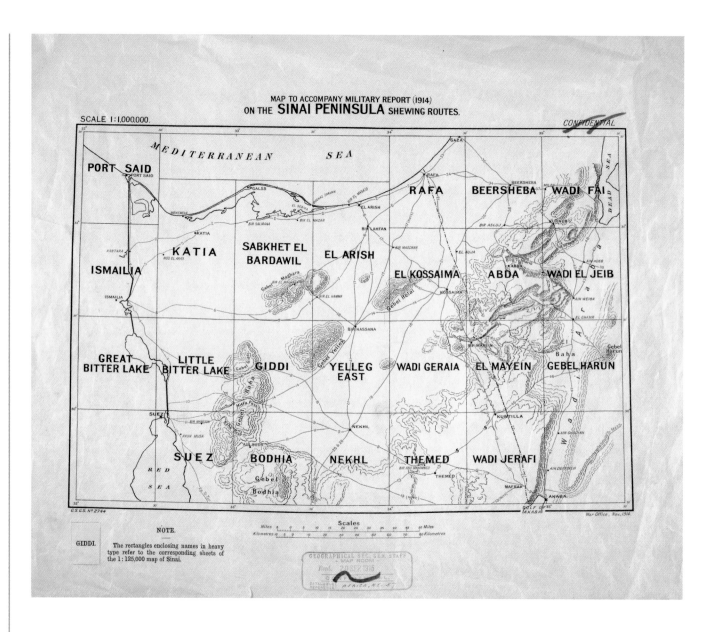

Map of
Area occupied by Australian & N.Z. Army Corps.
Contours at intervals of 40, Feet.
Heavy contours " " 200, "
Scale 1·10,000,

BATTLE OF GALLIPOLI, 1915 This map shows the area occupied by the Australian and New Zealand Army Corps (ANZAC). The landings were hindered by the extent to which the Turks had strengthened local defences under German direction, and by poor Allied generalship and planning. The initial assault was not pressed forward hard enough and this left commanding high ground to be quickly occupied by the Turks. Their fighting skills proved important, as was the general strength of defensive firepower. The fighting rapidly became static, with the tiny, exposed Allied positions made even grimmer by the heat, dysentery and, eventually, typhoid. These were factors that could not be shown on the map. In May, Lieutenant-General Sir William Birdwood, the ANZAC commander, reported that he had been pushed back onto the defensive and 'practically reduced to a state of siege', in the face of repeated Turkish attacks. Moreover, mobility had been lost, with Birdwood complaining in June: 'It seems quite ridiculous that we should be within ten yards of each other, and yet I am unable to get into their trenches.' LEFT

ANZAC PANORAMA
AUGUST 7ᵗʰ 1915
Note: The observer is standing on a high ridge with his back to the sea.
Right rear, Anzac Beach; left rear North Beach.

MAP OF GALLIPOLI PANORAMA ON 7 AUGUST 1915. BY CAPTAIN LESLIE FRASER STANDISH HORE It shows the dispositions of the Australian forces on the day of the attack on the ridge known as The Nek by the 8th and 10th Australian Light Horse Regiments. Published in Cairo in 1916, the notes provide much detail, including the arrows showing the Australian front line. A combination of poor command decisions and firm Turkish resistance led to failure in August, as also with the original fighting in April. Mobility was hard to regain. In May, John Monash, an Australian brigade commander, observed: 'We have been fighting now continuously for 22 days, all day and all night, and most of us think that absolutely the longest period during which there was absolutely no sound of gun or rifle fire, throughout the whole of that time, was ten seconds.' ABOVE

written by the Chaplain to the Forces, the atlas included an autograph portrait of Winston Churchill, First Lord of the Admiralty. On the inside page 'Signatures of the brave. A place for the autographs of officers and men who served Britain by land and sea in the Great War of 1914' topped Shakespeare's lines 'This happy breed of men … this England.'

Historical continuity was also demonstrated to the public in *The Oxford historical wall maps* (1915). This series included a map of the Low Countries in 1914, with an inset of the situation in 1702, when, at the outset of the War of the Spanish Succession, Britain had also deployed a major army to protect its interests there and defeat a hostile neighbouring power (in that case, ironically, France, which in 1915 was Britain's key ally). For both domestic and foreign audiences, maps for information and maps as propaganda were types that overlapped. They reflected the extent to which the wars were perceived as total struggles and, in particular, ones in which it was important to mobilise the support of the 'home front'. This was an aspect of the way in which conscription and war finance arose from a different social politics and political system to those of a century earlier. Mass literacy meant that war news was expected and delivered, and the prime form, newspapers, ensured that there was a need for new information every day. To go with this, new images were required. The

limitations of photography made maps correspondingly more significant.

This significance posed a challenge as the war did not change on a daily basis, and it was also difficult to establish how best to satisfy public interest. Innovations, which included film, did not take a cartographic form. In contrast to the development of precise aerial photography as part of an integrated air–artillery information system, there was no comparable innovation in the mapping of the war for the civilian population. The aerial-style mapping, associated in particular with Richard Edes Harrison, that was to be seen in the Second World War was not yet developed at that scale, although there were maps that adopted aerial perspectives, not least in order to display the interplay of air and land warfare. However, there was a familiarity in mapping for the public based on a continuity of the location mapping seen, for example, in the maps printed for the Boer War of 1899–1902 in southern Africa. There is no sign that this led to any dissatisfaction. Looked at differently, the strategic opportunities, options and problems of the war were not widely debated by the public as newspapers and the politicians were under tight control. Partly as a result, simple location maps served most public purposes. As the next two chapters indicate, the situation was very different with the major conflicts later in the twentieth century.

EXPLAINING THE FIGHTING: THE SITUATION IN FLANDERS, G.F. MORRELL, 1915 British depiction of the situation in Flanders by the autumn of 1915, elucidating a comment made in the House of Commons by Prime Minister Herbert Asquith. Gains made by Germans through use of gas at Ypres in April were countered by British advances in the Battles of Neuve Chapelle and Loos in April and September respectively. These battles revealed the defensive strength of trench systems, as did the French failure in Champagne in September. The concentration of large forces in a relatively small area ensured that any defender was able to call on plenty of reserve troops to stem an infantry advance. In each case, initial success was not exploited, in part due to command flaws but also because of primitive communications. Germany remained in control of the gains made in 1914, which was a key strategic context. Moreover, German successes against Russia greatly increased the need for the Allies to take steps on the Western Front to reduce the pressure on Russia. LEFT

PICTURE MAP TO ILLUSTRATE THE BATTLE OF NEUVE CHAPELLE.

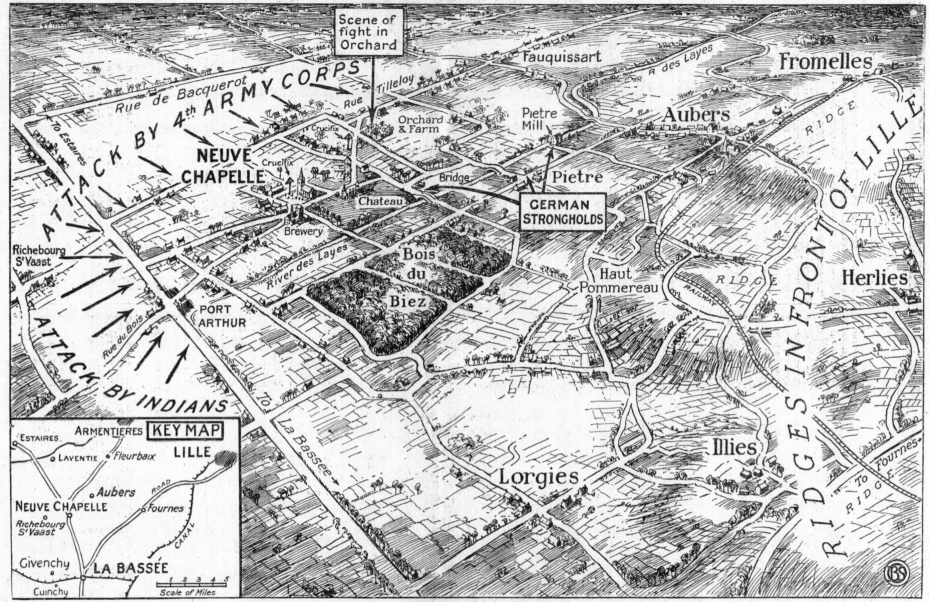

The large map indicates the approximate lines of advance of the British and Indian troops against Neuve Chapelle. The small map shows the position of Neuve Chapelle in relation to Lille, the great French manufacturing town held by the Germans.

BATTLE OF NEUVE CHAPELLE, 1915 This picture map illustrates the Battle of Neuve Chapelle, with a smaller inset to show the location of the battle in relation to Lille. At Neuve Chapelle on 10 April 1915, despite a weak bombardment that did not cut all the German wire, the British had sufficient mass to break through German trenches which did not yet display the sophistication they were to show later in the year. Frontal attacks could therefore still do well. Although the German defence was assisted by the British lack of heavy guns and ammunition, as well as by a British loss of control of the battle because of poor communications, the British nearly succeeded at Neuve Chapelle and, indeed, came closer than in many subsequent battles. However, the Germans quickly brought up reserves by train and established a fresh front line, for which the British artillery was entirely unprepared, while many of the British troops were exhausted. Success evaporated, and additional attacks failed. Private Stanley Green commented on the stunned nature of the surviving attackers and wrote that the operation proved very different to the 'visions of thrilling charges and hand to hand combat' he and others had anticipated. **ABOVE**

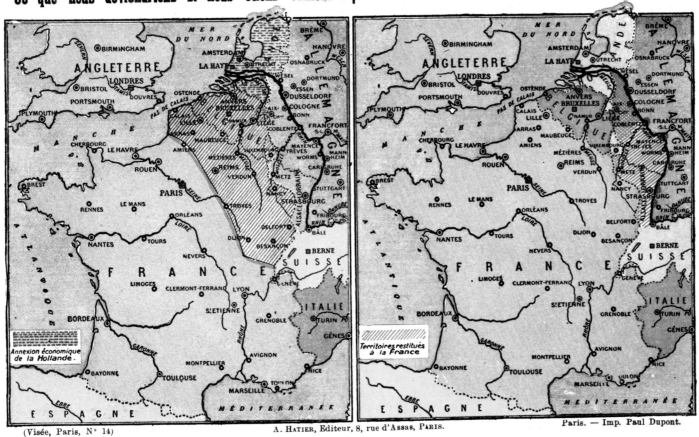

LES AMBITIONS ALLEMANDES | CE QUE VEULENT LES ALLIÉS

d'après les documents allemands | Pour la Paix de l'Europe

Ce que nous deviendrions si nous étions vaincus ! | Le Rhin frontière de l'Allemagne

(Visée, Paris, N° 14) — A. HATIER, Éditeur, 8, rue d'Assas, PARIS. — Paris. — Imp. Paul Dupont.

WARTIME PROPAGANDA POSTCARD, PRINTED BY THE FRENCH *LIGUE DES PATRIOTES*, 1915 The left-hand map indicated German goals, the right-hand one Allied goals allegedly focused on providing European peace. The official goals announced by Bethmann-Hollweg, the German Chancellor, on 9 September 1914 were less extensive but included the Longwy-Briey iron ore basin from France, dominance of Belgium and colonial gains in Africa. Moreover, naval commanders were keen on obtaining bases on the coast of Belgium and, if possible, France. **LEFT**

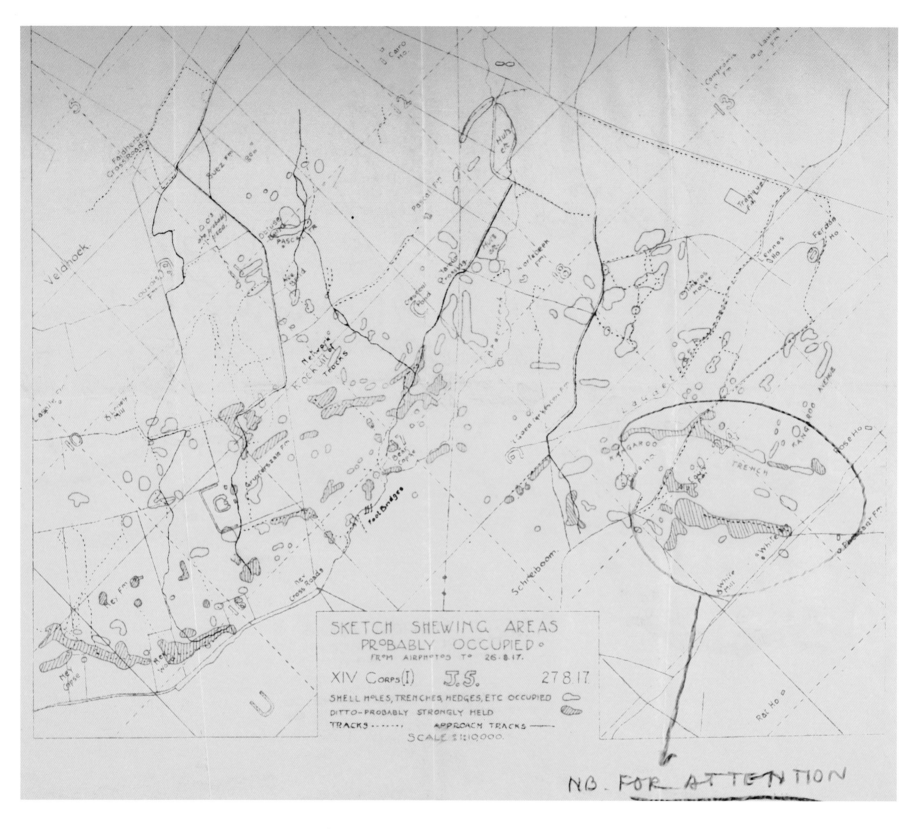

AERIAL RECONNAISSANCE, 1917 'Sketch showing areas probably occupied from Air photos of 26 August 1917.' British aerial reconnaissance of German positions on a section of the Western Front. A map that was clearly used as the underlined annotation 'For Attention' shows. Aircraft were useful for aerial mapping, although the correct method of reading maps taken from the air, based on aerial photography, was still in its infancy. The Germans had gained the advantage in the air in the winter of 1916–17 thanks to the entry in service in September 1916 of the aerodynamically efficient, fast, manoeuvrable Albatross D-1 which was armed with two synchronised machine guns. The Germans lost their aerial advantage from mid-1917 as more and better Allied aircraft entered service, especially the French Spad S.VII and Spad S.XII and the British Bristol Fighter, with its forward and rearward bearing armament and its synchronising gear based on wave pulses, and the extremely nimble Sopwith Camel. These aircraft offered the Allies a quantitative advantage, providing support for reconnaissance aircraft. **ABOVE**

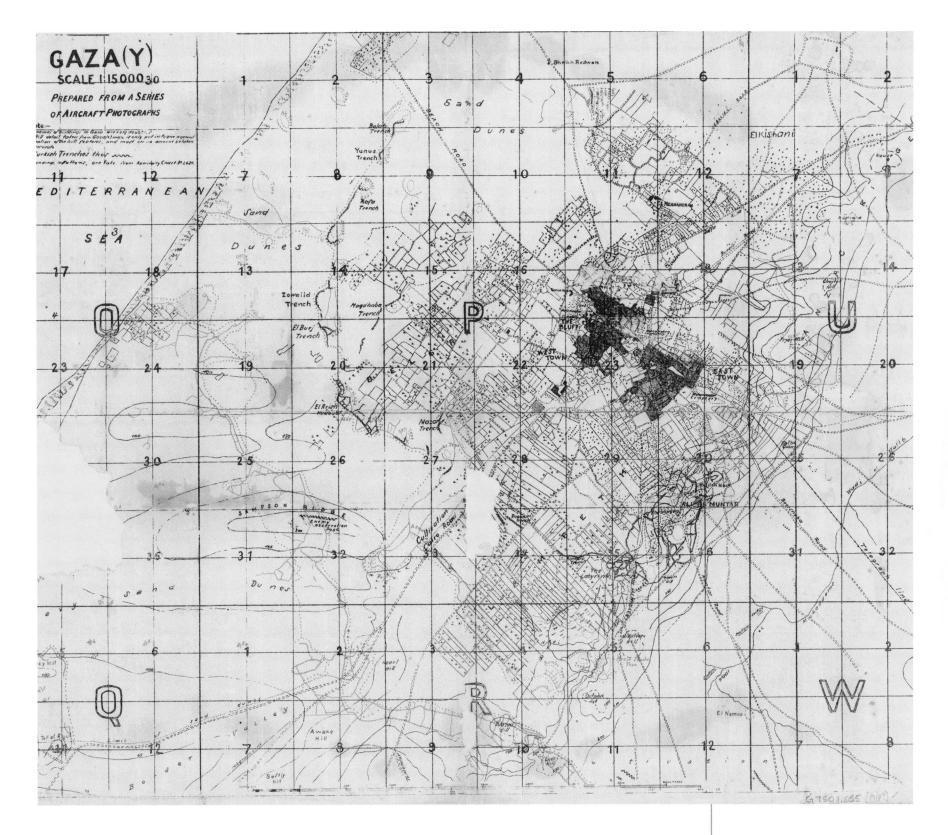

GAZA, 1917, THE SURVEY OF EGYPT Aerial photography was the basis for this map produced by the Survey of Egypt, the key British agency for mapping of the eastern Mediterranean littoral. After crude early attempts, a successful method of mapping from aerial photographs was developed in 1917, leading to the mapping of the entire frontal line from Gaza to Beersheba. In the spring, the British imperial forces launched two unsuccessful offensives against the well-defended Turkish lines at Gaza. In October, a third offensive was launched, with tactical lessons from the warfare on the Western Front employed to help break through the Turkish positions, while, further east, a surprise Australian attack captured Beersheba. ABOVE

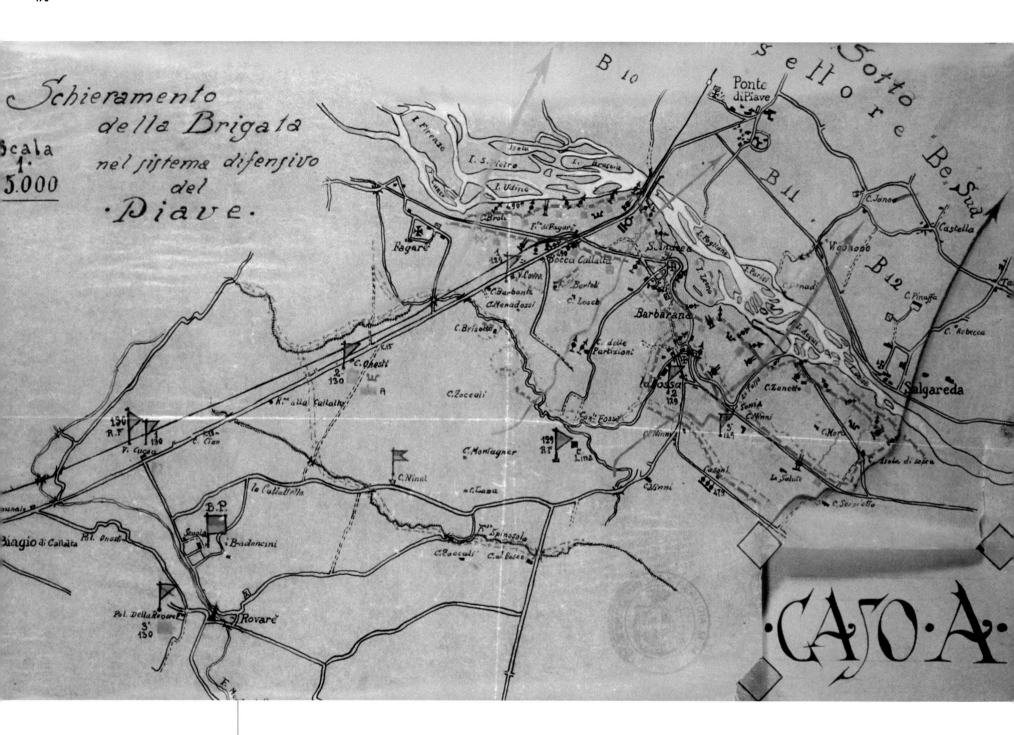

Schieramento della Brigata nel sistema difensivo del Piave

Scala 1:5.000

ITALIAN DEFENSIVE SYSTEM ON THE RIVER PIAVE, 1917
This map shows the deployment of the 20th Brigade in the Italian defensive system on the River Piave. After defeat by attacking Austrian and German forces at Capretto in October 1917, a new front line on the Piave River was shored up in November with major Allied contingents. The Austrians attacked on the Piave River front towards Padua in June 1918, but were beaten back with heavy losses. Drawing on their experience, the British had helped improve the effectiveness of the Italians in trench warfare, which was part of the process by which Italian warmaking improved during the war. The Austrians crossed the Piave, but could not secure their position on the western bank. Instead, under pressure from the impact of air attacks and a rain-swollen river on their supply routes, and affected by the Italian ability to deploy reserves, the Austrians withdrew their troops. The *Carta topografica d'Italia alla scala 1:25.000*, a national base map authorised in 1878, was created principally for military purposes. ABOVE

SECOND BATTLE OF THE MARNE, 1918 This American depiction of information from captured German maps, prisoners' statements and aerial photography illustrates the Second Battle of the Marne on 2 October 1918. The Americans copied British and French techniques, notably the use of aerial photography. German resistance continued to be strong as the Germans were pushed back in a series of Allied attacks along the front line. These Allied attacks were designed to wear them out and crumble their defences, making it impossible for them to consolidate a new front. Launched on 26 September, the American Meuse-Argonne offensive had taken heavy casualties as it fought its way through strong defences; and this costly progress remained the pattern when the reinforced offensive was resumed on 4 October, although the Germans were driven from the Argonne Forest. The map provides valuable information on German rear areas, information crucial for sustaining any breakthrough. On 2 October, emphasising the situation on the Western Front, the German Supreme Army Command, in a briefing to leading German politicians, made the need for an armistice apparent. **LEFT**

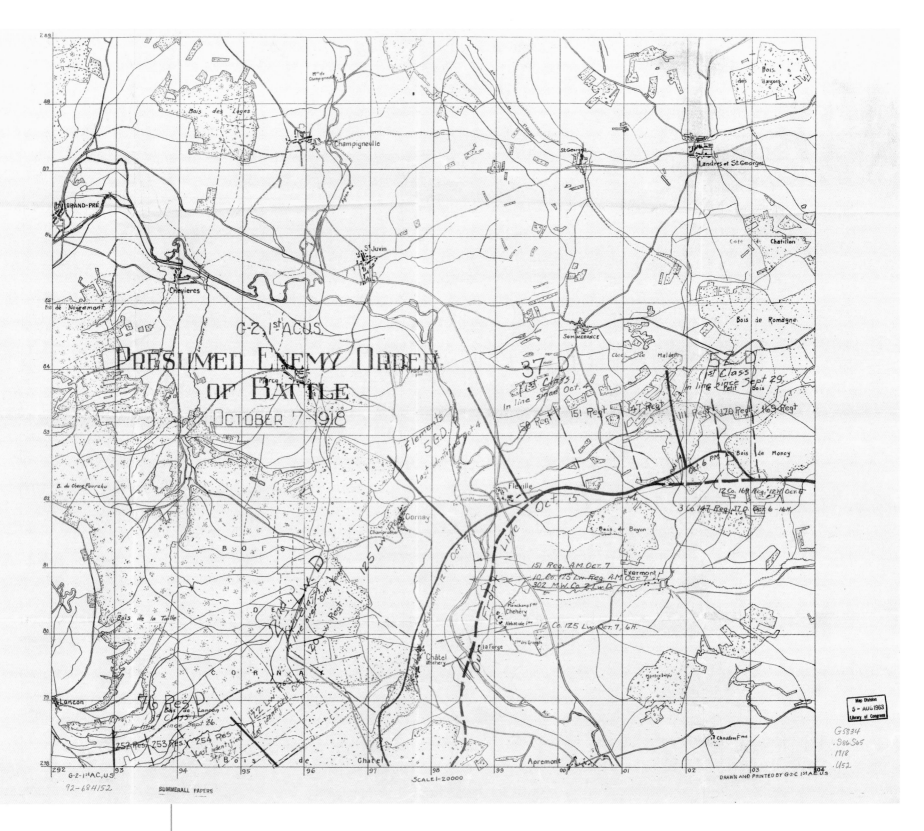

INFORMATION ON THE ENEMY, 1918 American map of 7 October 1918, providing information from captured German maps, prisoners' statements and recent aircraft photographs. Such maps provided indications of German defences in depth and thus guidance to the likely problems that would be encountered if the German front line was breached. The Americans were on the offensive in this period. Launched on 26 September, the American Meuse-Argonne offensive had taken heavy casualties as it fought its way through strong defences; and this costly progress remained the pattern when the reinforced offensive was resumed on 4 October, although the Germans were driven from the Argonne Forest. On 1 November, the last of the American offensives was launched as the Americans fought their way across the River Meuse. The fall of German positions suggested that none of their defensive positions were invulnerable. **ABOVE**

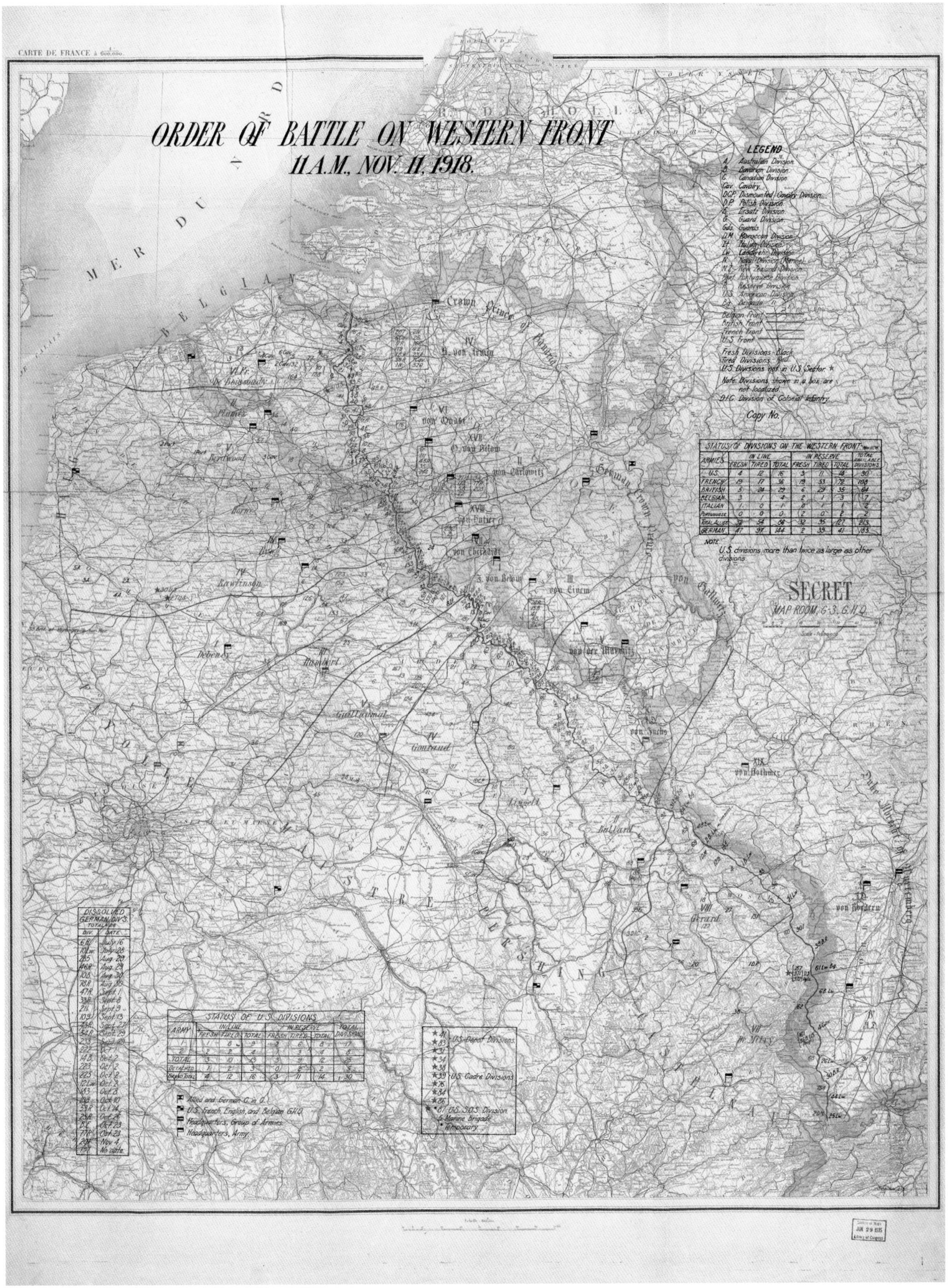

WESTERN FRONT DURING THE ARMISTICE, 1918

This map shows the order of battle on the Western Front on 11 November 1918 at 11 am, when the armistice came into force. While support for the war within Germany had collapsed, successive German defence lines on the Western Front were broken through. The correspondence of many Allied officers and soldiers reflected both confidence in their activities and a sense that the situation had changed abruptly. Not only had the Allies overcome the tactical problems of trench warfare, but they had also developed the mechanisms, notably greatly improved logistics, and deployed the resources, especially large numbers of guns, necessary to sustain their advance in the face of continued resistance and across a broad front. On 3 November, Den Fortescue of the British army noted 'what has regulated our pace much more than the Boche resistance is the difficulties of our own communications'. Fortescue also wrote that the Germans were heavily outgunned as many of their guns had been captured. Lieutenant-Colonel Alan Thomson wrote on 11 November that, having had a trumpeter sound the ceasefire, 'I then called for three cheers for King George and the many troops who had assembled in the courtyard responded right heartily. It was a stirring moment.' LEFT

GLOBAL CONFLICT

SHANGHAI, 1932 Map of the city of Shanghai during the Japanese attack of 1932, showing the French and international settlements and the Japanese and Chinese battle lines. The town of Chapei, which is close to the frontlines, is depicted with houses burning down. Main map is a bird's eye view from a low oblique angle. The Japanese attacked the Chinese section of Shanghai from late January 1932. After early hesitation, there was strong resistance with German-trained divisions fighting well against the Japanese in a large-scale conflict to which the Japanese eventually committed over 50,000 troops as well as 200 aircraft and warships. The devastation caused by Japanese bombardment and bombing was very great, while the landing of a Japanese amphibious force up the Yangzi River outmanoeuvred the Chinese who fell back from Shanghai. In the end, under international mediation, China accepted the demilitarization of Shanghai while Japanese troops were pulled back from Chinese areas of the city. **RIGHT**

SECOND WORLD WAR – D-DAY LANDINGS AT OMAHA BEACH, 6 JUNE 1944 Amid rough seas, the US V Corps landed tens of thousands of troops from a vast flotilla of battleships and other craft. **PREVIOUS PAGES**

IN THE PERIOD between the wars (1918–39), there were technological changes that were to affect mapping for war. In particular, there were further developments in aerial photography. In the 1930s, these developments included the introduction of colour film and infrared film. Infrared images can present colours otherwise invisible to the eye, and infrared film can distinguish readily between camouflaged metal, which does not reflect strongly in the infrared colours, and the surrounding vegetation, which does. Black-and-white film, however, retained a value as the contrast was stronger than with colour film and therefore more helpful.

Preparing for large-scale war led to innovations with mapping, notably with the British development of radar. This reflected the conviction that bombing might determine the course of the next war between major powers. Radar was a reflection of the new nature of the three-dimensional character of mapping that stemmed from the addition of aircraft to the vertical space represented by terrain. The likely significance of air power made this a major factor as a defensive capability. As with the use of mapping for artillery, radar was a response to the need to fix position accurately. However, unlike artillery, aircraft posed an inherently dynamic character in location and, thus, the depiction of location. Radar, thereby, looked ahead to what became a key element in the depiction of the battle space.

Maps also served to mobilise people for struggle, with the Nazi regime encouraging a depiction of the German past in terms of struggle with outsiders, although this theme was already present in atlases published prior to the Nazi takeover in 1933.

The wars of the interwar period did not see much mapping unless they involved states with well-developed mapping systems, notably Britain and France. Both actively mapped new possessions, notably in the Middle East and particularly when they were faced with rebellions, as the British were in Iraq (1920) and Palestine (1936–39) and the French in Syria (1925–26).

In contrast, insurrectionary forces did not have comparable facilities for mapping. For example, in China, the Communist People's Liberation Army (PLA) could not carry out the systematic and professional military survey necessary to produce the maps they required for military operations in the late 1920s and, even more, the 1930s. Instead, the PLA used civilian maps with information added and made copies of captured Guomindang (Nationalist) military maps.

UNPRECEDENTED SCALE

In the Second World War (1939–45), the production and use of maps were even greater than in the First World War. The contrasting factor of scale was readily apparent. The British Ordnance Survey produced about 300 million maps for the Allied war effort, and the American Army Map Service (AMS) produced more than 500 million. In addition, there were numerous other military mapping organisations producing large numbers of maps. Production techniques responded accordingly.

Shortages were rapidly overcome. For example, a shortage of maps in military collections when the United States entered the war in December 1941 necessitated extensive government borrowings, notably from the New York Public Library and an appeal to the public for maps:

The War Department, Army Map Service, is seeking maps, city plans, port plans, place name lexicons, gazetteers, guide books, geographic journals, and geologic

NORTH

TSUNGMING ISLAND 崇明嶼

YANGTSE RIVER 揚子江

PICTORIAL MAP
OF
SHANGHAI
AND ITS ENVIRONS, INCLUDING A SKETCH MAP OF
THE ACTUAL FIGHTING ZONE AT TAZANG, CHENJU
AND KIANGWAN SHOWING THE LARGER VILLAGES

上海市鳥瞰畫及附近村區圖

KEY:-
RED SHADING = International Settlement
BLUE SHADING = French Concession
Black line thus - - - - means:- The Japanese front line for over one month
Purple line thus - - - "   The Chinese front line for over one month
Red line thus - - - "   The Japanese new point line after defeating

NANZIANG means:- New Head quarters of Japanese army.
RIFLE RANGE "   :- Original Head quarters "
CHENJU "   :- Chinese "   until they were defeated

P.S. Those white marks in Chapei & Chung means fire. 20,000
houses were burnt in Chapei alone.

This drawing has been prepared by
E. B. CUMINE, A.R.I.B.A., A.A.dip.
from Maps issued from reliable
sources and every effort has been
made to use popular or newspaper
spelling for proper names appearing

ERIC CUMINE

SOUTH

SINO-JAPANESE WAR, 1937   Map showing the development of the war and the confrontational relationship between Japan and the Soviet Union. Published as a supplement to the November 1937 issue of a Japanese women's magazine, *Fujin Kurabu*. The map shows the front line of the Sino-Japanese conflict as of 1 October 1937, as well as places bombed by the Japanese, and the disposition of Soviet troops. Charts include the development of the Sino-Japanese War and the armaments of the Soviet army. Korea was then a Japanese colony. Manchuria in northern China had been occupied by the Japanese in 1931-2. RIGHT

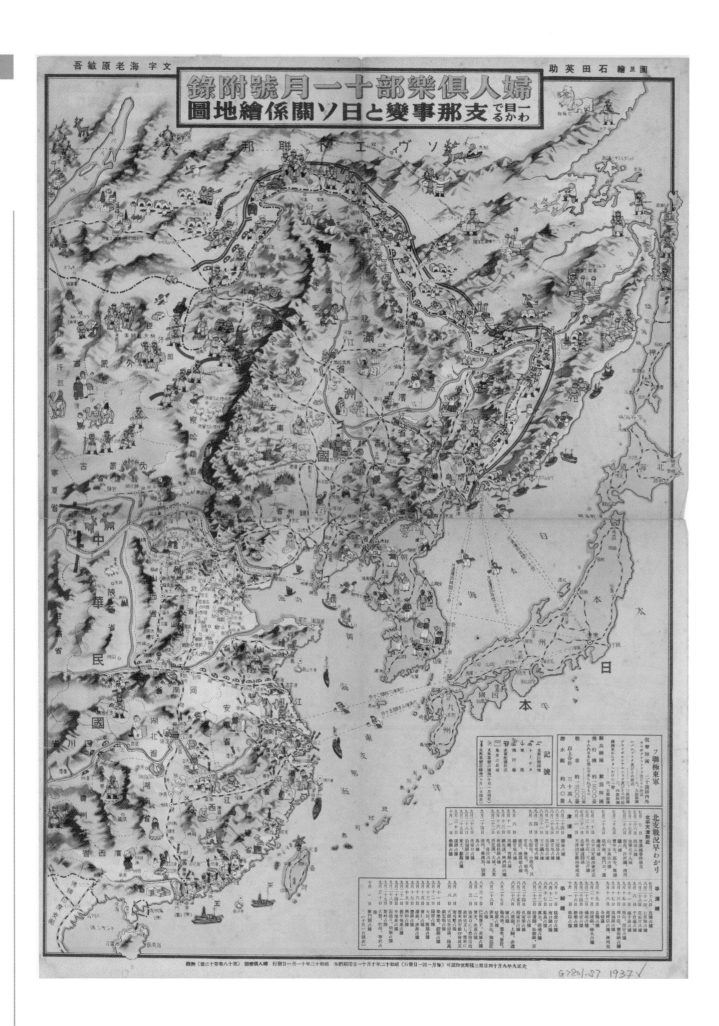

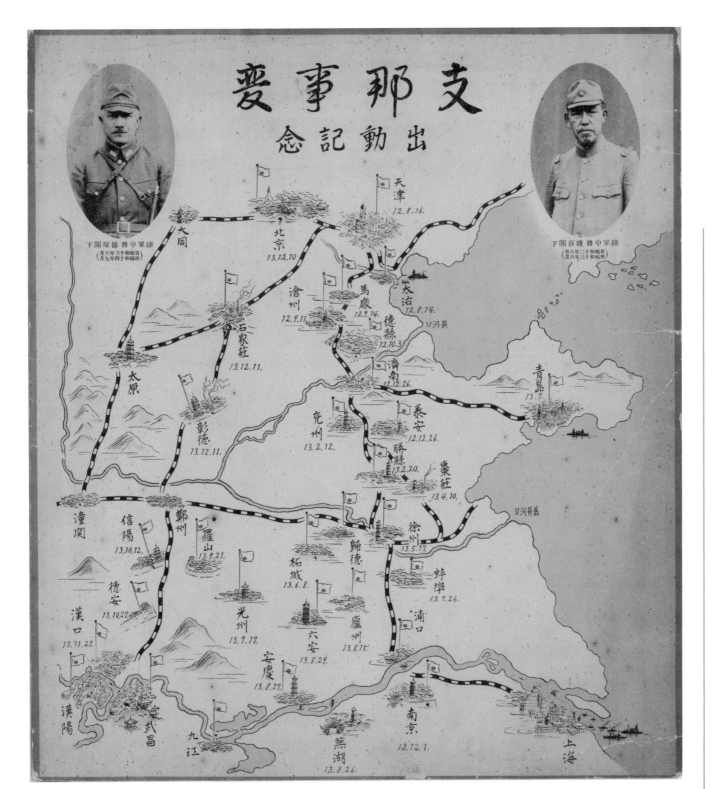

支那事變
出動記念

陸軍中將能塚壉下
（昭和十四年六月三十日）
（昭和十三年六月九日）

陸軍中將磯谷下
（昭和十三年六月三十日）
（昭和十三年六月二十日）

THE SINO-JAPANESE WAR, 1937–1938 Map in commemoration of Japanese advance into central China, 1937–38. The capture of Beijing, Shanghai and Nanjing are all covered. The inset photographs are of successive commanders of the 10th Division. Annotated in Japanese on the reverse, 'Kishimoto Takeo, a 1st-class private, who gave this map to Kunitoshi on June 26, 1940'. Beijing fell rapidly to the Japanese in July 1937, but the Chinese proved far stronger in the Guomindang heartland of the lower Yangzi valley. Shanghai resisted from August to November, only falling after bitter, house-to-house, fighting. The Japanese Central China Expeditionary Force then advanced on the Guomindang capital, Nanjing, which fell on 13 December after a heavy air attack and the breaching of the city walls by artillery. The subsequent slaughter of large numbers of civilians in part was a deliberate step to crush resistance. LEFT

bulletins covering all foreign areas outside the continental limits of the United States and Canada. Of particular interest are maps and guide books purchased within the last ten years.

This situation was swiftly remedied. Units in the field were not only provided with plentiful maps, but were also able to produce them in response to new opportunities and problems. These included the defence of the United States against possible attack. Spitsbergen was important due to its asbestos, coal and gypsum mines and as a source of crucial meteorological information.

As a result of such needs, cartographic expertise was at a premium. Walter Ristow from the map division of the New York Public Library became head of the

POSTER CRITICAL OF THE BRITISH GOVERNMENT'S
ATTITUDE IN THE SPANISH CIVIL WAR, BY JAMES
FRANCIS HORRABIN, 1938   An example of
maps being used as propaganda, this
British poster criticises the attitude of
the British government in the Spanish
Civil War. The threat of Spain to British
and French trade routes is highlighted.
The map-maker, James Francis Horrabin,
was a Labour MP, member of the
National Council of the Socialist League
and critic of British imperialism.   RIGHT

THERE IS GRAVE DANGER

that the Government will, during the next week or so

ACTUALLY HELP THE FASCISTS TO STARVE THE SPANISH PEOPLE INTO SURRENDER

This would happen if they agreed to the **Granting of Belligerent Rights to Franco**, which he desires in order to enable him to blockade Spanish Government ports.

DO YOU THINK IT JUST—after the Spanish Government has held out year after year with incredible heroism, and is now sending away all its foreign volunteers, in spite of the attacks from all the troops and aeroplanes Germany and Italy have still got in Spain—that

We, Great Britain, should now help to defeat them by starving their children?

Our Foreign Secretary, Lord Halifax, has said in the House of Lords that " Mussolini has always made it plain from the time of the first conversations with the British Government, that for reasons known to all of us, whether we approve them or not, he was not prepared to see General Franco defeated."

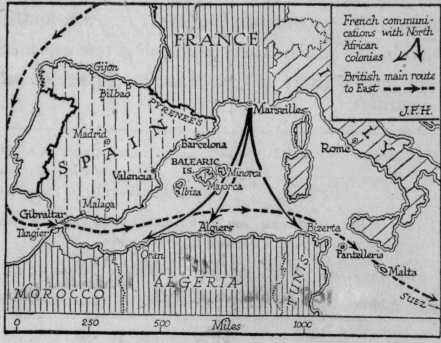

Map by J. F. Horrabin.

P.T.O.

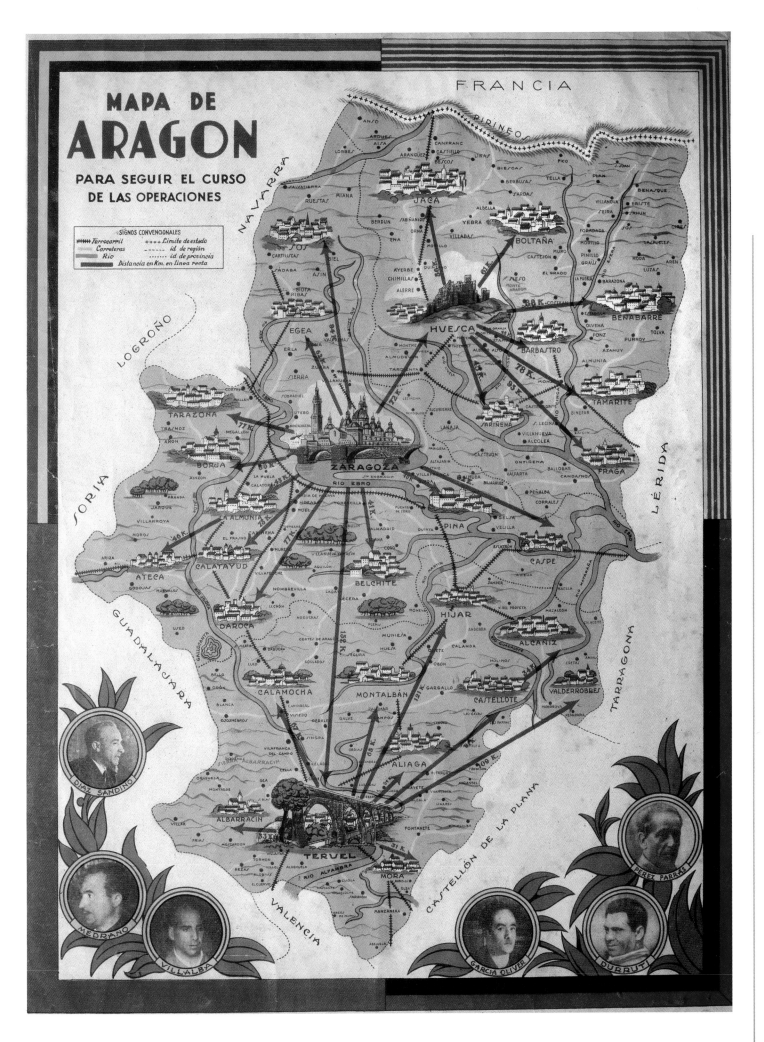

OPERATIONS IN ARAGON, 1936 The Spanish
Civil War (1936–39) arose as the result of
an attempted coup against the left-wing
government of Spain launched by a group
of right-wing army officers, who called
themselves the Nationalists and who felt
threatened by a Soviet-style revolution.
Instead of achieving a rapid success, as
was the case with many coups, the
Nationalists met strong resistance. Their
partial success led to a long, bitter and
ideological civil war. This map depicted
Republican success in the early stages of
the Spanish Civil War. The map was
produced by the Republicans showing
the civil war in the province of Aragon in
the opening stages. Aragon was one of
the few areas where Republican loyalists
took back territory, and the map shows
loyalist advances from the major cities
into the surrounding countryside. A sense
of success is clearly communicated. The
colours round the map include the flags
of the Republic, Catalonia and the
Anarchist movement. The last was
prominent in Aragon and its militias
played a key role there in 1936.
Photographs of Anarchist leaders are
included. **LEFT**

THE GERMAN CAMPAIGN IN POLAND, SEPTEMBER

1939 An illustration from the book *Die Soldaten des Führers im Felde* published in 1940. In a rapid German ground offensive that began on 1 September, the cohesion of the Polish army was destroyed. German armoured forces broke through, isolated, and enveloped the Polish formations. A poorly executed Soviet invasion of eastern Poland from 17 September, in cooperation with the Germans, completed the picture of Polish vulnerability. Warsaw surrendered to the Germans on 27 September and the last of the troops stopped fighting on 6 October. The Germans had killed 70,000 Polish troops and taken 694,000 prisoners. Fewer than 15,000 German troops were killed. **RIGHT**

geography and map section of the New York Office of Military Intelligence. Armin Lobeck, professor of geology at Columbia University, who had written a major study of geomorphology in 1939, produced maps and diagrams in preparation for Operation Torch, the successful American invasion of French North Africa in 1942. Lobeck also produced a set of strategic maps for Europe. Specific maps and atlases were also requested, for example, in June 1942, the *Atlas of Svalbard* (1927).

THREE DIMENSIONS

The need to coordinate air and land, air and sea, and land and sea operations ensured the increased complexity of many maps, especially to help in planning in three dimensions. This was true of bombing and tactical ground support from the air and of airborne attacks by parachutists and gliders. The development of the operational dimension of offensives, particularly in the Soviet army in 1943–45, notably with Operation Bagration in 1944, required an ability to seize and retain the initiative and to

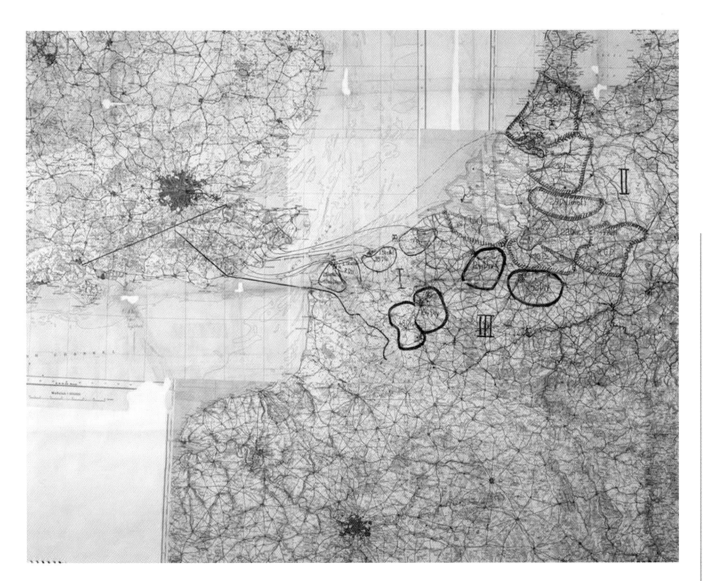

outmanoeuvre opponents. This ability heavily depended on staff work that was well informed about the distances and locations involved and maps were critical to the planning.

Similarly, long-range bombing (for example by the Americans of Japan from 1942 and, more consistently, 1944) required particularly accurate maps because much of the bombing was at, or close to, the limit of the bombers' flying range. Having acquired the *Atlas of Japan* (1931) produced by the Japanese Land Survey Department, as well as many other maps of Japan, the American Office of Military Intelligence had them reproduced. It also created a visual index to the atlas plates using a 1937 railway map of Japan with a hand-drawn grid showing the locations of the individual plates.

More generally, to help bombers, existing printed maps were acquired and supplemented by the products of photo-reconnaissance and other surveillance

activity. To guide their bombers to targets in Britain, the Germans used British Ordnance Survey maps enhanced with information from photo-reconnaissance. Maps and sophisticated map-linked radio navigation systems were developed to help night flying. The navigator was a key figure in bomber crews.

The use of aerial photography greatly expanded and improved, with better cameras offering more magnification. Equally significant, the analysis of the film by photo interpreters improved. Photo-reconnaissance was important for military planning, as with the German attack on the Soviet Union in 1941. This was preceded by long-range reconnaissance missions by high-flying aircraft. The Allied invasion of Normandy in June 1944 was preceded by more than two years of aerial photography, and the resulting maps were highly detailed. They also benefited from an understanding of the coastal geology. The map-

MAPPING FOR BOMBING, BY *LUFTWAFFE*, 1941

Focusing on Liverpool, this was one of a
series used by German bombers in
night-time missions over Britain. Only the
most obvious features are included on
the map: towns, railway lines, major
roads, woodland etc. The maps' bright
colours helped the pilots to pick out
details at night. Areas marked in red and
yellow, the target areas, reflected the
focus on attacking dock areas, which
complemented the submarine assault in
the attempt to starve Britain into
surrender, the key policy for the
Luftwaffe in late 1940 and early 1941, and
one that helped explain the air assault on
London's dockland. The yellow circle
surrounding the area just outside Oldham
alerted the bombers to a prisoner of war
camp holding German soldiers. Aerial
photography was a major source for the
Luftwaffe. British night-time air defence
techniques were poor; not least because
of the absence of precise aerial radar, it
was hard to hit German aircraft once
they largely switched to night attacks in
September 1940 in part in order to cut
losses. The German bombers mostly got
through. Relatively few German aircraft
were shot down that winter, although
accidents and the winter weather led to
many losses. By the late spring of 1941,
however, the British defence was
becoming more effective thanks to the
use of radar-directed Beaufighter planes
and ground defences. RIGHT

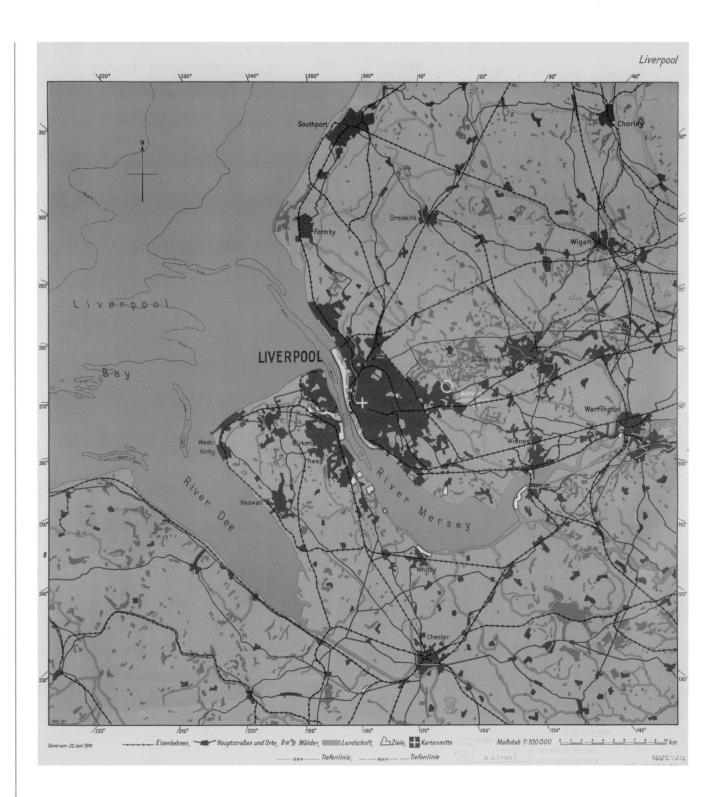

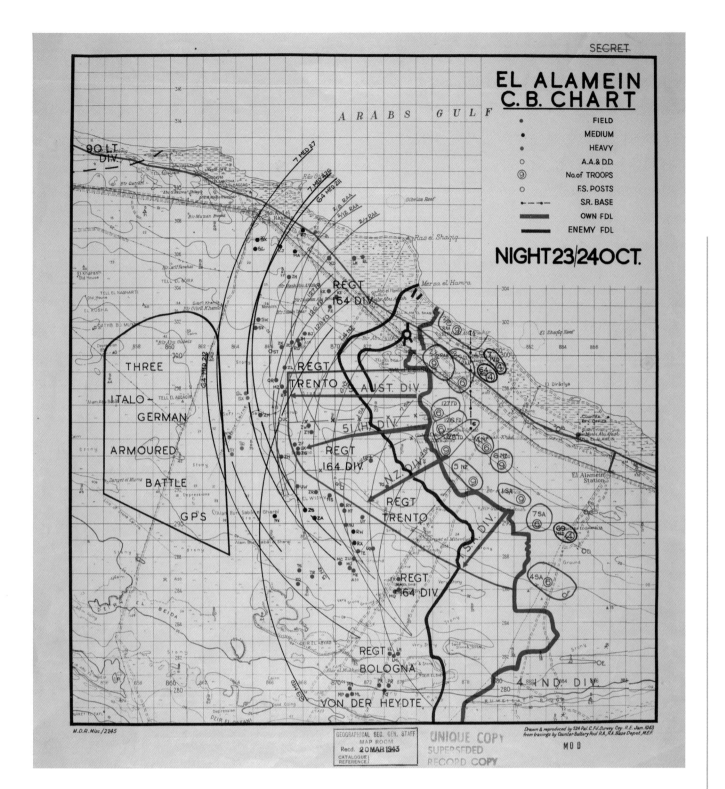

OPERATION LIGHTFOOT. BY ROYAL ENGINEERS

PALESTINIAN SURVEY COMPANY, 1942

Launched on the night of 23–24 October
1942, the British plan for the second
battle of El Alamein, Operation Lightfoot,
was a deliberate, staged assault launched
against static defences without any
turning of an open flank. General Bernard
Montgomery planned to breach the
northern part of the German-Italian
defences by attacking on a ten-mile
front, clearing routes through the
minefields so that the British armour
could advance. In the event, there were
serious problems causing delay. However,
eventually, skilful generalship, greater
numbers of troops, artillery and tanks,
effective use of artillery and air
superiority ensured that sequential blows
finally succeeded, with the Axis line
broken on 4 November. In cartographic
terms, the British benefited from
operating in an area they had controlled
and surveyed prior to the war, with the
Survey of Egypt producing valuable data.
During the war, military surveyors in
Egypt used planes, and aerial
photography was important for mapping
behind enemy lines. In 1942, maps of the
suitability for traffic of the terrain were
produced. LEFT

makers drew on the work of special units which
mapped the beaches, mostly at night. These maps had
to include tidal levels and coastal defences.

Maps were more extensively used for ground
operations than in the First World War. In part, this
was because of the greater mobility of units, not least
on the Western and Eastern Fronts, in Italy and at sea.
Reliable maps and mapping skill encouraged planning
to reach a specific point on a map at a certain time,

planning already seen in the First World War, but now
linked to greater mobility.

In addition, the use of maps within the military was
more widely extended. Whereas ordinary soldiers
(unlike officers and senior non-commissioned officers)
did not use maps extensively during the First World
War, they did so in the Second World War. Given the
extent to which armies were mass forces raised by
conscription, this meant that knowledge of map use

BEKANNTMACHUNG

betr. die Bildung eines geschlossenen jüdischen Wohnbezirks in der Stadt Lublin.

OBWIESZCZENIE

Dotyczy: Utworzenia w Lublinie zamkniętej, żydowskiej dzielnicy mieszkaniowej.

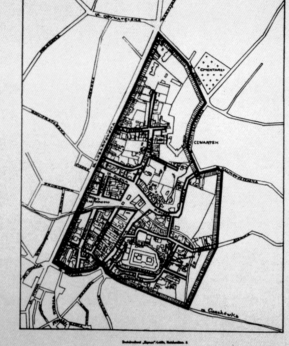

spread widely in society; this had already happened to a degree with driving, notably in the USA thanks to the use of the *Rand McNally Road Atlas* in the 1920s and 1930s.

The improvements in mapping took a variety of forms. For example, to encourage accurate use, reliability diagrams were widely introduced on military maps.

RESTRICTIONS AND THE MEDIA

Secrecy has always been crucial to mapping for war, and this was also true in the Second World War. Aside from producing maps, governments restricted the distribution of maps that might help enemies; in the United States, this applied to topographic maps and nautical charts. The lack of available information greatly enhanced the value of capturing maps. In 1941, the Germans benefited in the early stages of Operation Barbarossa, their invasion of the Soviet Union, from the capture of the Minsk map factory, followed by those in Kiev and Kharkov.

In turn, the Soviet Union moved fast to complete the 1:100,000 map of the state west of the River Volga, a task that entailed 200 new sheets and was completed by the end of 1941. Moreover, the Red Army developed a flexible production system in which maps were prepared and printed in order to meet immediate operational needs. This entailed the overprinting of maps with specific tactical information. Underpinning this activity was a more general attempt to improve the training for map-making and the production of maps, including the introduction of a uniform geodetic system.

As in the First World War, maps were again used both to provide news and for propaganda. Newspapers printed large numbers of maps. The newspaper

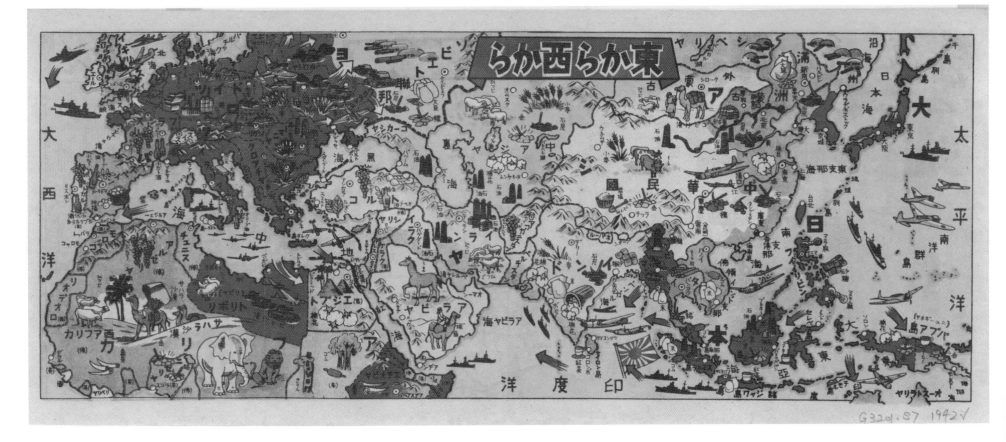

らか西らか東

G320.87 1942V

mapping of the war subordinated all spatial considerations to the front line. Indeed, the key element was the location of the line and, linked to that, the employment of additional lines and shading to display the change in the front line and the use of arrows to show the direction of attacks. This approach ensured an operational focus in the cartographic presentation of the war. Tactical-scale details and/or examples were not generally presented in newspaper maps, while the strategic dimension, though advanced, received less attention than its operational counterpart.

The maps introduced readers to areas about which they knew little, notably, for the British and American public, the Eastern Front in the Soviet Union. Indeed, the idea that war news should be accompanied by maps owed much to the world wars and particularly the Second World War. This represented a form of visual information very different to that of newsreels and the television, but one that helped prepare the way for aspects of the latter.

However, the notion of a front line was singularly inappropriate in some areas. This was especially true of the Pacific where a number of important Japanese bases, such as Rabaul and Truk in the south-west Pacific, were leapfrogged or bypassed in 1943–45

thanks to superior American air and sea power. This situation could not be readily captured on the maps. Similarly, the impact of autumn rains, winter freeze and spring thaws on the roads on the Eastern Front in 1941–45 were not apparent on the maps.

Prominent propaganda maps included the Vichy map showing Britain as a Churchillian octopus reaching out to attack the French empire and being repelled. This particular map drew on French suspicion of British global activity and was an aspect of the more general use of maps in propaganda.

Again atlases were produced, including the *Atlas of the war* (Oxford, 1939), *The war in maps* (London, 1940), and *The war in maps: an atlas of the New York Times maps* (New York, 1943). The aerial dimension encouraged the use of particular perspectives and projections for maps.

ENGAGING THE AMERICAN PUBLIC

The innovative cartographer Richard Edes Harrison had introduced the perspective map to American journalism in 1935. His orthographic projections and aerial perspectives brought together the United States and distant regions, and were part of a worldwide extension of American geopolitical concern and

THE AXIS ADVANCE, 1942 Japanese pictorial map of Europe, Asia and northern Africa, shows areas of Axis occupation. Map of Pacific region on the verso has areas of Japanese occupation, American bases, locations of sinking of enemy ships and defence lines against the enemy. Map published as 1942 edition of a Japanese annual magazine that was presented as easy to read and fun. The map shown captures the potential threat to the Allied position in South Asia presented by the German advances toward Egypt and the Caucasus and the Japanese advance in the Indian Ocean region. However, the threat was also exaggerated. The Italians in Eritrea had been defeated in April 1941, those in northern Ethiopia, the remaining Italian forces in East Africa, following in May–November 1941. Moreover, the Japanese raid into the Bay of Bengal in April 1942 was not sustained and did not, as shown, continue into the Arabian Sea. Nor did Crete serve as a base for German attacks on Egypt. ABOVE

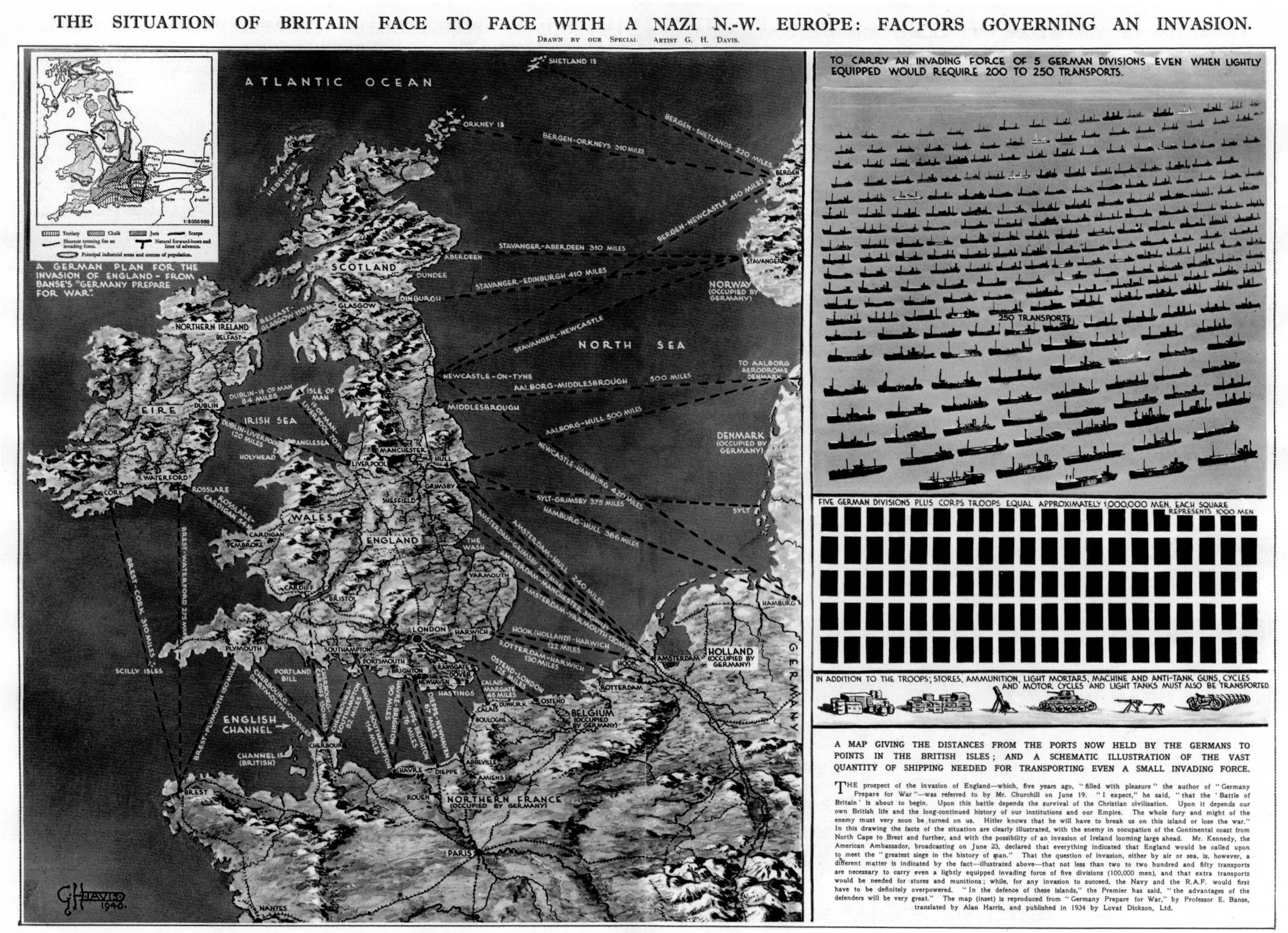

THE SITUATION OF BRITAIN FACE TO FACE WITH A NAZI N.-W. EUROPE: FACTORS GOVERNING AN INVASION.

DRAWN BY OUR SPECIAL ARTIST G. H. DAVIS.

A MAP GIVING THE DISTANCES FROM THE PORTS NOW HELD BY THE GERMANS TO POINTS IN THE BRITISH ISLES; AND A SCHEMATIC ILLUSTRATION OF THE VAST QUANTITY OF SHIPPING NEEDED FOR TRANSPORTING EVEN A SMALL INVADING FORCE.

THE prospect of the invasion of England—which, five years ago, "filled with pleasure" the author of "Germany Prepare for War"—was referred to by Mr. Churchill on June 19. "I expect," he said, "that the 'Battle of Britain' is about to begin. Upon this battle depends the survival of the Christian civilisation. Upon it depends our own British life and the long-continued history of our institutions and our Empire. The whole fury and might of the enemy must very soon be turned on us. Hitler knows that he will have to break us on this island or lose the war." In this drawing the facts of the situation are clearly illustrated, with the enemy in occupation of the Continental coast from North Cape to Brest and further, and with the possibility of an invasion of Ireland looming large ahead. Mr. Kennedy, the American Ambassador, broadcasting on June 23, declared that everything indicated that England would be called upon to meet the "greatest siege in the history of man." That the question of invasion, either by air or sea, is, however, a different matter is indicated by the fact—illustrated above—that not less than two to two hundred and fifty transports are necessary to carry even a lightly equipped invading force of five divisions (100,000 men), and that extra transports would be needed for stores and munitions; while, for any invasion to succeed, the Navy and the R.A.F. would first have to be definitely overpowered. "In the defence of these islands," the Premier has said, "the advantages of the defenders will be very great." The map (inset) is reproduced from "Germany Prepare for War," by Professor E. Banse, translated by Alan Harris, and published in 1934 by Lovat Dickson, Ltd.

'THE SITUATION OF BRITAIN FACE TO FACE WITH A NAZI N.W. EUROPE: FACTORS GOVERNING AN INVASION', **FROM** *THE ILLUSTRATED LONDON NEWS*, **1940** This map, published in *The Illustrated London News* on 29 June 1940, gives distances from the ports held by the Germans to points in the British Isles. It includes a schematic illustration of the vast quantity of shipping required for transporting even a small invasion force. Hitler's hope for a negotiated settlement

initially took precedence and on 19 July he offered peace, with Britain to retain her empire. Winston Churchill was not interested in negotiating. On 16 July, Hitler had ordered preparations for an invasion, but the German navy was concerned about a lack of capacity and demanded air supremacy, a point also stressed by Hitler. The losses suffered by the German navy during the Norway campaign meant that its ability, in the face of British

naval power, to support an invasion had been lessened. Moreover, the *Luftwaffe* could only try to prevent British naval interference in daylight, allowing any invasion fleet that could not cross the Channel and return in daylight to be attacked at night. In practice, aside from a failure to gain air supremacy, neither the navy nor the air force were committed to a Channel crossing while, allowing for the invasion of Norway earlier in the year, the army did not

have adequate experience, resources or interest. The text argues that the Germans faced formidable transport requirements for any invasion and would first need to overcome the Royal Navy and the Royal Air Force. The inset map was reproduced from Ewald Banse's *Germany, prepare for war!*, the 1934 translation of his *Raum und Volk im Weltkriege* (1932). **ABOVE**

GERMAN PROPAGANDA, FROM THE FRENCH EDITION OF *SIGNAL*, 1941 This map emphasises the challenge of German submarines to British trade and warships and of German aircraft to British cities. The German conquests of Norway and France in 1940 exposed more of British trade to submarine attack than in the First World War; while submarines were more sophisticated than in that conflict and the submarine assault on British trade unrestricted from the outset. However, the Luftwaffe totally failed to devote sufficient resources, in aircraft or aviation fuel, to the Battle of the Atlantic against British shipping. In the winter of 1940–41, the German air attacks focused on London and in March to April 1941 on the British ports, but in May the bombers were moved to prepare for the invasion of the Soviet Union launched in June. **LEFT**

military intervention. Indeed, the role of aircraft, dramatically demonstrated to the Americans by the Japanese surprise attack on Pearl Harbor in December 1941, led to a new sense of space, a sense that reflected both vulnerability and the awareness of new geopolitical relationships.

The Esso war map (1942), for example, emphasised the wartime value of petroleum products, with 'transportation – key to victory' the theme of the text. It also provided an illuminated section, 'flattening the globe', which showed how the globe becomes a map. North America was placed central to the map, which made Germany and Japan appear as menaces from east and west. The map included sea and air distances between strategic points such as San Francisco and Honolulu. Public engagement was a clear theme.

President Franklin Delano Roosevelt's radio speech to the nation on 23 February 1942 made reference to a map of the world in order to explain American strategy. He had earlier suggested that potential

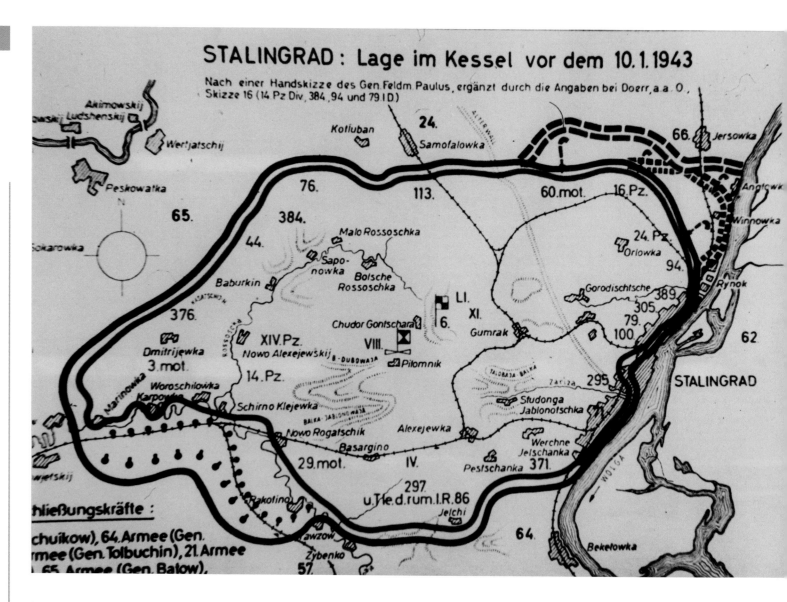

STALINGRAD : Lage im Kessel vor dem 10.1.1943
Nach einer Handskizze des Gen. Feldm. Paulus, ergänzt durch die Angaben bei Doerr, a.a.O., Skizze 16 (14. Pz. Div., 384, 94 und 79. I.D.)

GENERAL MAP OF THE STALINGRAD BASIN, SHOWING THE FRONT AS AT DECEMBER 1942 TO 10 JANUARY 1943, BASED ON A SKETCH BY GERMAN COMMANDER-IN-CHIEF GENERAL PAULUS
The German 6th Army and parts of the 4th Panzer Army, with support from Hungarian and Romanian forces, fought the Soviet army for control of the city of the southern Soviet city of Stalingrad for more than five months from late August 1942 to February 1943. Their ground attack was supported by *Luftwaffe* bombing that destroyed much of the city. This map shows the front towards the end of the campaign, after the Soviet counteroffensive had cut off the German forces. The campaign, one of the most decisive of the whole war, was a severe defeat for the German army. German losses were so heavy that they were obliged to take reinforcements from western Europe to replenish their forces, seriously weakening their position elsewhere in Europe. RIGHT

listeners obtain such a map, which led to massive demand and increased newspaper publication of maps. The previous year, Roosevelt had claimed to possess a secret map of German designs on Latin America, one that he knew had been forged by the British.

Already on 9 September 1939, Rand McNally had announced that more maps had been sold at its New York store in the first 24 hours of the war than during all the years since 1918, as public interest and concern rose greatly. There was another upsurge in sales after America entered the war following Pearl Harbor. Roosevelt, who obtained his maps from the National Geographic Society, created a map room in the White House. For Christmas 1942, he was given, by George Marshall, the Chief of Staff, a huge and detailed fifty-inch globe that had been specially made for him under

the supervision of the map division of the Office of Strategic Services and the War Department.

The task of explaining engagement with distant regions, and of throwing light on these regions, posed a problem, but also produced opportunities for innovation in both conception and presentation. The American film *Man for Destruction* (1943) depicted the actor playing the German geopolitician Karl Haushofer, who was presented as close to the Nazi regime, explaining global geopolitics in front of a map centred on the North Pole. This was an exposition of a threat that linked different parts of the world and suggested that such a map helped explain what the American response should be – an approach that looked toward the Cold War. Richard Edes Harrison also produced a map centred on the North Pole, with

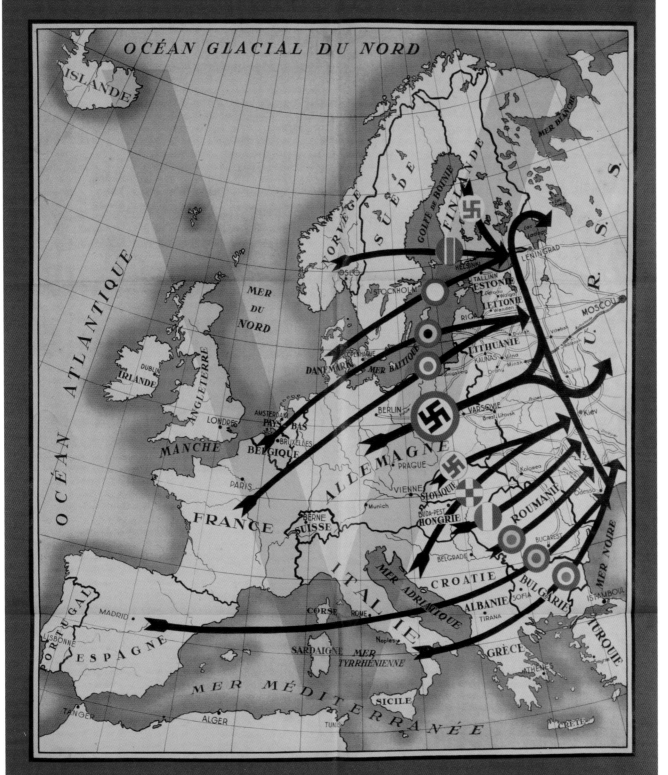

WARTIME PROPAGANDA POSTER, BY 1941 This 1941 poster about a united European crusade against Communism is aimed at a French audience. The presentation of Operation Barbarossa was designed to help foster and direct European enthusiasm for the German cause. Allies and volunteers did indeed provide valuable support, notably Finns and Romanians on the German flanks. The theme of a crusade was in part designed to draw on Catholic sympathy and a number of prominent Catholic clerics and laity were willing to adopt this position. The absence of any depiction of a Franco–German border serves to avoid underlining France's defeat the previous year. The map suggests that the advance to Moscow would be easy. LEFT

ALLIED AIR OFFENSIVE AGAINST GERMANY UP TO 1 JANUARY 1941 A poster designed to show that not only British cities were being bombed. However, the bombing of Germany, notably of the cities of Berlin and Mannhein, in late 1940 was on a smaller scale and inflicted much less damage. Industrial sites in the Ruhr and naval bases were also attacked. The RAF, in practice, lacked not only the number of aircraft but also an effective bomber. The small bombers, such as the Hampden I and the Blenheim, were not fast enough and had poor defensive capabilities. Indeed, the concept of the small bomber was not fully thought out. These aircraft were withdrawn from bombing in 1942 and 1943 respectively. Almost half of the Hampden Is that were built had been lost. Moreover, Stirling and Halifax bombers lacked the power and bomb load of the four-engined Lancaster. The German conquest of the Netherlands, Belgium and France in 1940 created many difficulties for British bombers attacking Germany. The need to provide fighter escorts was not properly appreciated. The figures next to each target depicted the number of major attacks. The black lines represented number of miles from London. RIGHT

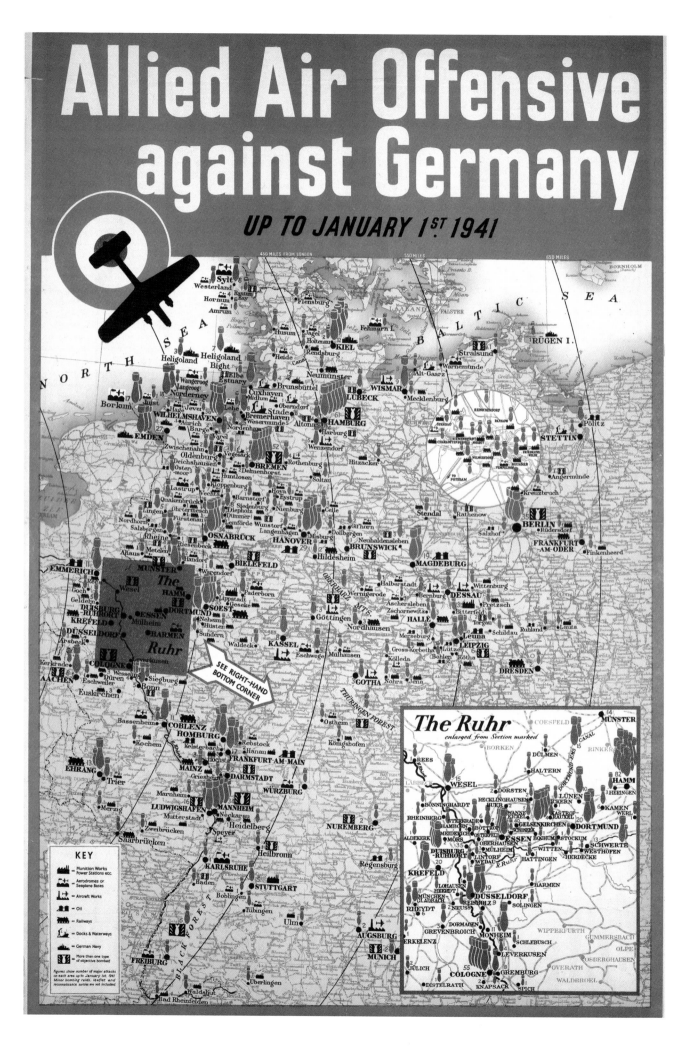

BRITAIN – SPEARHEAD OF ATTACK POSTER, 1942 This poster captured Britain's strategic position, notably as the main air base for Allied air attacks and as a major manufacturing centre. Air attacks built up in 1942, with the Lancaster (bomb load 22,000lb) entering operational service in March. Most of the raids in 1942 were fairly small scale, with the cities of the Ruhr industrial region, notably Essen and Dortmund, the main targets. The raids, however, were important to the development of an effective ground support system to underpin a bombing offensive, as well as in the gathering of operational experience. More than 1,050 bombers were launched against Cologne on the night of 30–31 May 1942. The bomber stream tactic was used, where bombers gathered in one stream in order to use mass to counteract the power of the defences. The caption notes that the navy assures 'to Britain a ceaseless and gigantic flow of supplies'. **ABOVE**

THE CRIMEAN OFFENSIVE, 1944 This Soviet poster celebrates the defeat of the Germans in April and May 1944. The boldness in size, solidity, colour and direction of the arrows used to show the Soviet forces contrast with those for the Germans. As the Soviets had the initiative, they could choose when and where to attack, and the isolated German and allied Romanian units in Crimea played no real role until the moment of destruction. The conquest of Crimea took just over a month. Soviet propaganda trumpeted their ability to capture Sevastopol more speedily than the Germans had done in 1942. Hitler had ignored Romanian requests for withdrawal from Crimea, and had proved all too receptive to assurances that first Crimea and then Sevastopol could be held. RIGHT

the USA presented in a key position.

The preface to Harrison's *Look at the World: the Fortune Atlas for World Strategy* (1944), an atlas that reproduced his maps from the magazine *Fortune*, explained that it was intended 'to show *why* Americans are fighting in strange places and *why* trade follows its various routes. They [the maps] emphasise the geographical basis of world strategy.' Harrison's maps put the physical environment before national boundaries, and also reintroduced a spherical dimension, offering an aerial perspective that does not exist in nature, but that captured physical relationships, as in his 'Europe from the southwest', 'Russia from the south', 'Japan from Alaska,' and 'Japan from the Solomons'. The first edition of the atlas rapidly sold out, while Harrison's techniques were widely copied.

The reporting and presentation of war, notably the dynamic appearance of many war maps, for example those in *Fortune*, *Life* and *Time*, with their arrows and

general sense of movement, helped to make geopolitics present and urgent. Far from the war appearing to American readers as a static entity and at a distance, it was seen as in flux. The maps also made the war seem able to encompass the spectator both visually, through images of movement and also, in practice, by spreading in his or her direction. The orthographic projection used for the map 'The Aleutians: vital in North Pacific strategy', published in *The New York Times* on 16 May 1943, depicted the island chain as the centre in a span stretching from China to San Francisco. This presentation made their potential strategic importance readily apparent, which was necessary as public support was limited for what appeared to be a minor operation. The Office of War Information followed up with *A War Atlas for Americans* (1944), which offered perspective maps. The dynamic possibilities of cartography were far more to the fore than in the previous world war and the results were vivid.

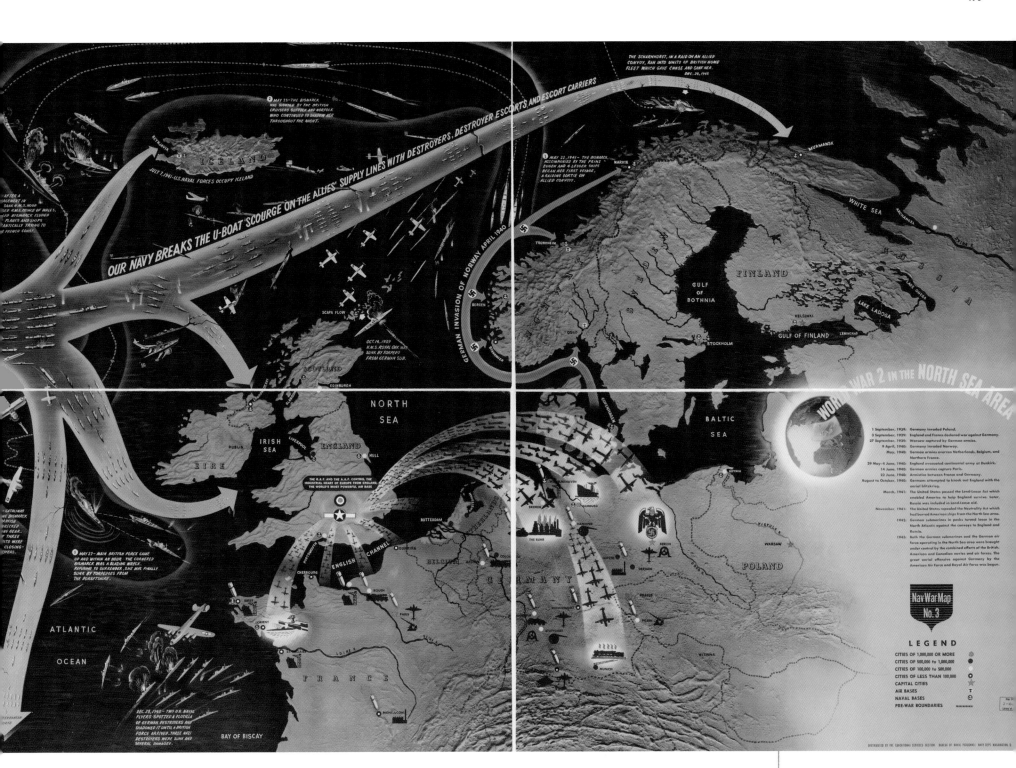

AMERICAN NAVAL WAR MAP. BUREAU OF NAVAL PERSONNEL, 1943 This map explained the course of the war to American naval personnel and was distributed by the educational services section of the Bureau of Naval Personnel. The aerial view carefully joins naval and air operations and British and American operations. The role of the Americans is somewhat exaggerated, but not excessively so, and due attention is devoted to the British. ABOVE

THE WAR CHANGES, 1944 Map of Europe showing Allied advances to the Australian public, Sydney's *Sunday Sun*, 10 September 1944. Mons had fallen on 2 September, Lyons on 3 September, and Florence on 4 August, although the Soviet failure to sen d help ensured that the Warsaw Rising against German control was being brutally suppressed at this time. The map did not capture the scale and extent of the Allied air attack, nor the strength of the Allied blockade. The map helps explain why it was widely believed that Germany could be defeated that year, a view that underrated the ability of the Germans to retain cohesion and maintain resistance as they retreated. The British military commentator J.F.C. Fuller claimed in the *Sunday Pictorial* of 10 September that 'at the moment, supply and not fighting power is the key factor of our advance,' and, indeed, all the Allies faced serious logistical problems. RIGHT

SUNDAY SUN

Week end MAGAZINE

SPECIAL SUPPLEMENT

SUNDAY, SEPTEMBER 10, 1944

Wresting Europe from the Nazis: This four color map shows graphically how the Allies on all fronts are rescuing Europe from its Nazi conquerors. Areas in red mark the Allied progress, yellow the areas still held by the enemy, green the neutral countries. Swift advances have altered the whole shape of the war in a few dramatic months.

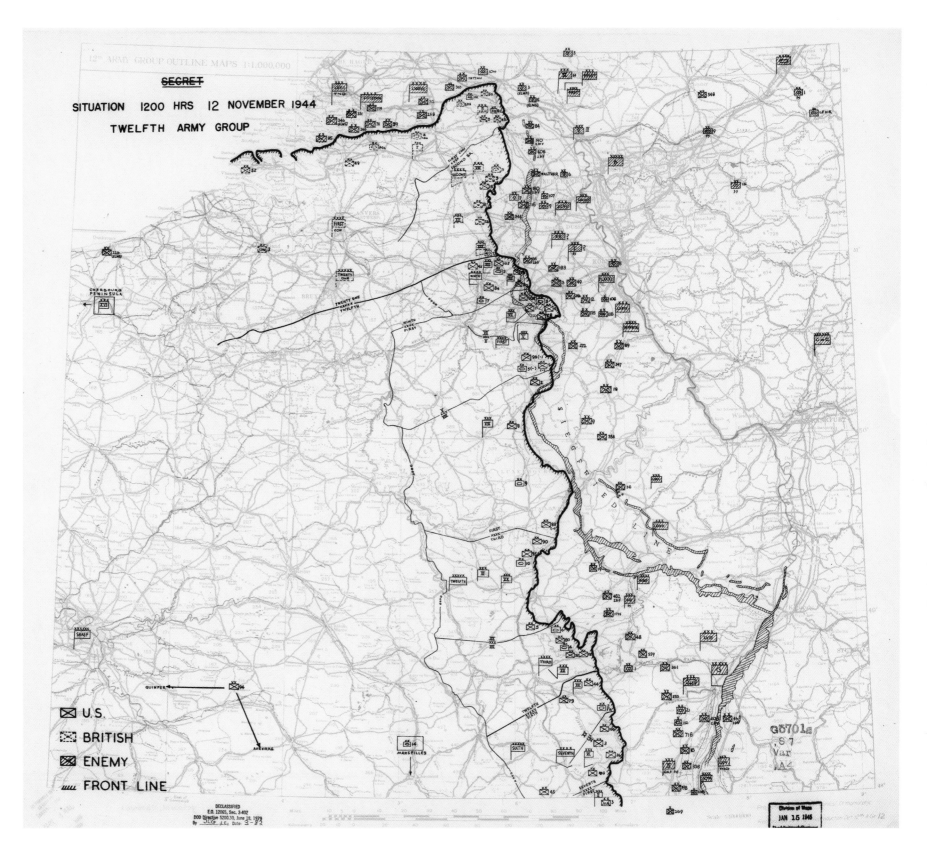

THE 1944 CAMPAIGN IN FRANCE. TWELFTH ARMY GROUP, 1944 This map shows the Twelfth US Army Group situation on 12 November 1944. Commanded by General Omar Bradley, this army group had played a key role in the advance across France, but had then stalled, notably in a slow attempt to clear the Hürtgen Forest. This attempt led to heavy casualties for the First US Army. The fighting quality the Germans showed there and elsewhere indicated the extent to which their earlier defeat in Normandy had not destroyed the German war effort. Allied hopes that the war would end that year proved premature. The strength of the resistance accentuated the debate over Allied strategy and operational choices. Logistical factors and exhaustion affected the Allies, while German supply lines shortened as they retreated. However, the 1944 campaign also showed that American, British and Canadian military capability, effectiveness and fighting quality had all improved greatly. **ABOVE**

TOKYO AND VICINITY

Showing bombed-out areas

For use by
Allied Forces personnel only.
Not for sale or distribution.

THE AIR ASSAULT ON TOKYO, 1945 1946 map showing the bombed-out areas of Tokyo which had been hit hard in 1945. Initially, the raids were long-distance and unsupported by fighter cover. This led to American attacks from a high altitude, which reduced their effectiveness. The raids were also hindered by poor weather, especially cloudy conditions, by strong tail winds, and by difficulties with the B-29's reliability including engine fires and other motor malfunctions. These, however, were sufficiently corrected to make it possible for the Americans to revise and carry out the massive bombing campaigns against Japan. From February 1945, there was a switch to low-altitude night-time area bombing of Japanese cities. The impact was devastating, not least because many Japanese dwellings were made of timber and paper, and burned readily when bombarded with incendiaries, and also because population density in the cities was high. Fighters based on the recently-conquered island of Iwo Jima (three hours by air from Tokyo) from 7 April 1945 could provide cover for the B-29s, which had been bombing Japan from bases on the more distant island of Saipan since November 1944. These attacks were designed to hit Japanese industrial production, in part by devastating the cities where much of it was based.

Alongside the attempt to mount precision attacks, the industrial working class was the target. Weaknesses in Japanese anti-aircraft defences, both aircraft and guns, eased the American task and made it possible to increase the payload of the B-29s by removing their guns. Although the Japanese had developed some impressive interceptor fighters, especially the Mitsubishi A6M5 and the NIK2-J Shiden, they were unable to produce many. **ABOVE**

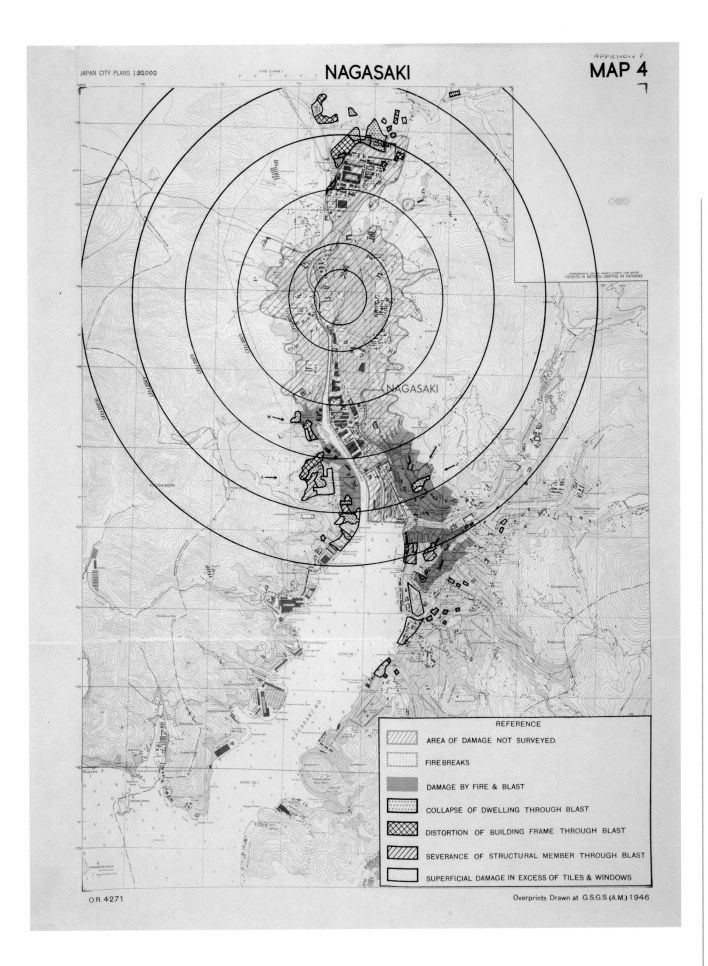

NAGASAKI

JAPAN CITY PLANS 1:20,000

APPENDIX F

MAP 4

REFERENCE

	AREA OF DAMAGE NOT SURVEYED.
	FIRE BREAKS
	DAMAGE BY FIRE & BLAST
	COLLAPSE OF DWELLING THROUGH BLAST
	DISTORTION OF BUILDING FRAME THROUGH BLAST
	SEVERANCE OF STRUCTURAL MEMBER THROUGH BLAST
	SUPERFICIAL DAMAGE IN EXCESS OF TILES & WINDOWS

O.R. 4271

Overprints Drawn at G.S.G.S (A.M.) 1946

NAGASAKI BOMB DAMAGE, BY ATOMIC WEAPONS ESTABLISHMENT, 1946 Map from a 1946 report of the British mission to Japan on an investigation of the effects of the atomic bomb dropped on Hiroshima and Nagasaki. Nagasaki was the target of the second atomic bomb, dropped on 9 August 1945. The combined shock of the two bombs led the Japanese to surrender. The Nagasaki bomb had a greater impact than the first one; the army had refused to believe what had happened and still wanted to fight after Hiroshima. The limited American ability to deploy more bombs speedily was not appreciated. **LEFT**

FROM THE SECOND WORLD WAR
to the PRESENT

CENTRAL ASIA, BY CIA, 1958 This terrain map of Central Asia was prepared using Swiss 'ridge and valley' techniques. The base terrain maps (shaped relief and lowland tints) were drawn on boards created from two sheets of paper fused to an aluminium core. The boards were accompanied by manuscript overlaps. These maps demonstrate sophisticated manual terrain mapping techniques now replaced by computer-generated terrain renderings using digital detail. **RIGHT**

FIRST GULF WAR, 24 FEBRUARY 1991 French troops in armoured vehicles enter Iraq through the desert. They were protecting the left flank of the US attack and swiftly overcame Iraq's 45th Infantry Division, suffering light casualties and taking many prisoners. **PREVIOUS PAGES**

THE COLD WAR, the struggle between the Communist and Western blocs, was consistently understood and presented in geopolitical terms, both for analysis and for rhetoric. As during the Second World War, a sense of geopolitical challenge was used to encourage support for a posture of readiness, indeed of immediate readiness. The sense of threat was expressed in map form, with both the United States and the Soviet Union depicting themselves as surrounded and threatened by the alliance systems, military plans and subversive activities of their opponents. These themes could be seen clearly not only in government publications, but also in those of other organisations. The dominant role of the state helps to explain this close alignment in the case of the Soviet Union and its Communist allies. In the United States, there was also a close correspondence between governmental views and those propagated in the private sector, not least in the print media.

PROJECTION

A sense of threat was apparent in the standard map projection employed in the United States. The Van der Grinten projection, invented in 1898, continued the Mercator projection's practice of exaggerating the size of the temperate latitudes. Thus, Greenland, Alaska, Canada, and the Soviet Union appeared larger than they were in reality. This projection was used by the National Geographic Society of America from 1922 to 1988, and their maps were the staple of American educational institutions, the basis of maps used by newspapers and television and the acme of public cartography. In this projection, a large Soviet Union appeared menacing, a threat to the whole of Eurasia and a dominant presence that required containment.

However, before employing these examples simply to decry American views then, it is necessary to point out that Soviet expansionism was a serious threat and that the geopolitical challenge from the Soviet Union was particularly acute due to its being both a European and an Asian power. This situation was captured by the standard Western depiction of the Soviet Union.

In turn, the Soviets employed a clearly hostile cartographic imagery and language, one that made the United States appear menacing and active. In short, although the 'roll-back' of Soviet power as a policy had been rejected by the Americans in the early 1950s (partly in response to the Soviet Union acquiring atomic bomb capability in 1949), the Soviets continued to depict the Americans as though such an aggressive policy was both their intention and means.

A sense of menace was repeatedly presented in cartographic form by both sides. Carrying forward President Roosevelt's use of maps to support his fireside chats over the radio, President John F Kennedy, in a press conference on 23 March 1961, employed maps when he focused on the situation in Laos, a French colony until 1954, where, in 1961, the Soviet- and North Vietnamese-backed Pathet Lao were advancing against the forces of the conservative government:

These three maps show the area of effective Communist domination as it was last August, with the colored portions up on the right-hand corner being the areas held and dominated by the Communists at that time. And now next, in December of 1960, three months ago, the red area having expanded – and now from December 20 to the present date, near the end of March, the Communists control a much wider section of the country.

The use of the colour red dramatised the threat, as did

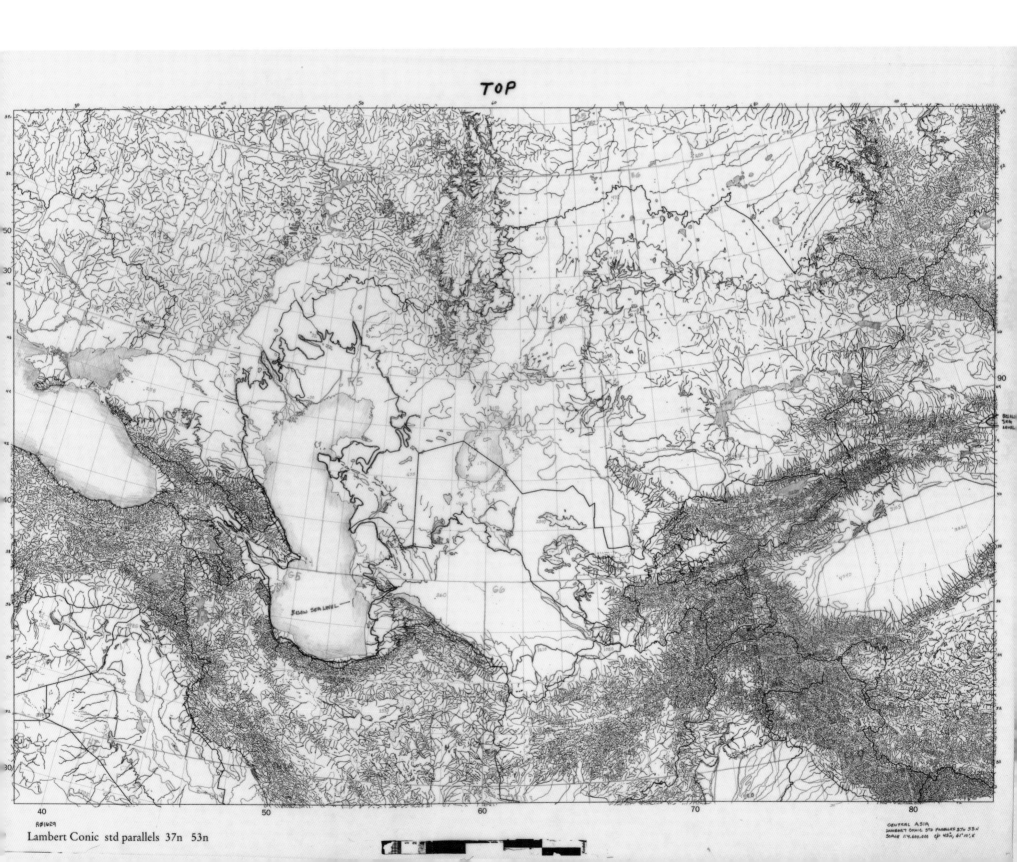

TOP

R01029

Lambert Conic std parallels 37n 53n

AMERICAN-BACKED INVASION OF CUBA, 1961

The newly-established Castro regime, itself the product of an insurrection, faced opposition as it pushed through a socio-economic revolution, notably the nationalisation of assets. However, in April 1961, President John F. Kennedy's failure to provide the necessary air support to a force of 1,300 CIA-trained anti-Communist exiles was blamed for the total defeat of their heavily-outnumbered invasion at the Bay of Pigs. In practice, poor planning and stiff opposition were also highly significant factors. When the exiles landed on 17 April, they met damaging air attacks, while the next day bombers sent to open the way for the landing of the necessary supplies for the stranded invasion force were mostly shot down. The supplies never arrived and, under heavy pressure from government forces, the rebels surrendered. RIGHT

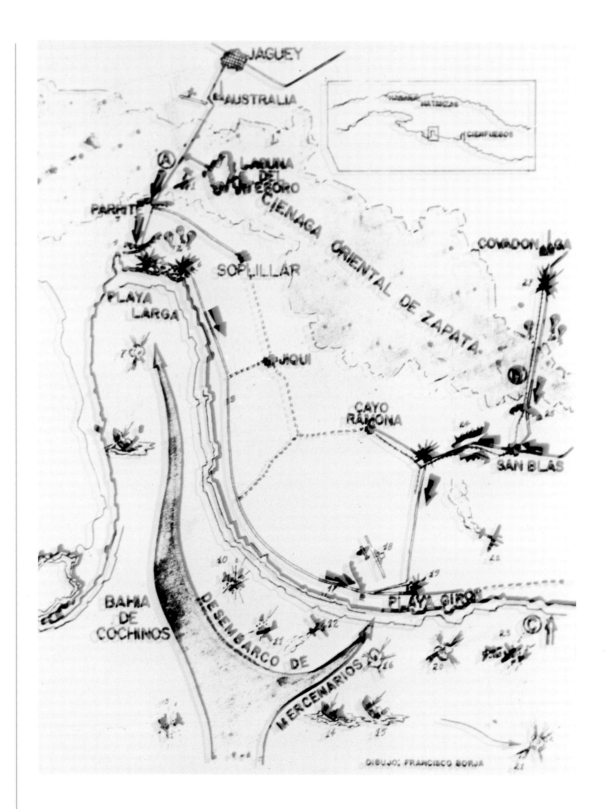

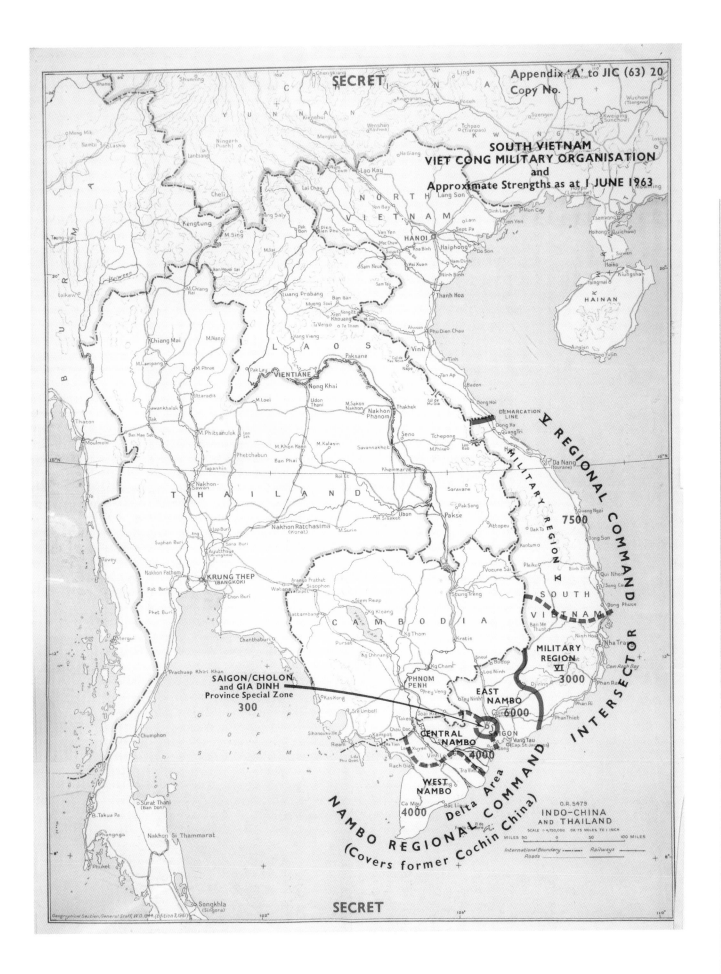

Appendix 'A' to JIC (63) 20
Copy No.

SOUTH VIETNAM
VIET CONG MILITARY ORGANISATION
and
Approximate Strengths as at 1 JUNE 1963

VIET CONG MILITARY ORGANISATION AND APPROXIMATE STRENGTH, BY BRITISH JOINT INTELLIGENCE COMMITTEE, 1963 British memorandum produced by the Joint Intelligence Committee in 1963. Under the Geneva Agreement terms, all-Vietnam elections were supposed to be held in 1956, but they were not, largely because of the opposition by the South Vietnamese government (led from 1954 to 1963 by Ngo Dinh Diem), which was brutal, corrupt and unpopular. The government had won the 1956 election in South Vietnam using fraud, and it represented the landowning élite that composed it. From 1957, South Vietnam faced a Communist rebellion by the Viet Cong, which resulted in more overt and widespread American intervention. From 1959, forces from North Vietnam were infiltrated into South Vietnam in support of the Viet Cong, who were linked to the North Vietnamese Politburo. LEFT

VIETNAM, MAPPING THE WAR, 1971   Map of Phuóc Tuy Province, Vietnam, overprinted on base map sheet 6430 from the Special Use S.E. Asia 1:100,000 topographical map series L607, covering the area around the Australian base at Nui Dat and overprinted to show sector and patrol boundaries, fire support bases, land clearance, artillery warning control centre boundaries, civilian access and coastal restriction areas, and amended place names. Produced in 1971 by the Royal Australian Survey Corps.   RIGHT

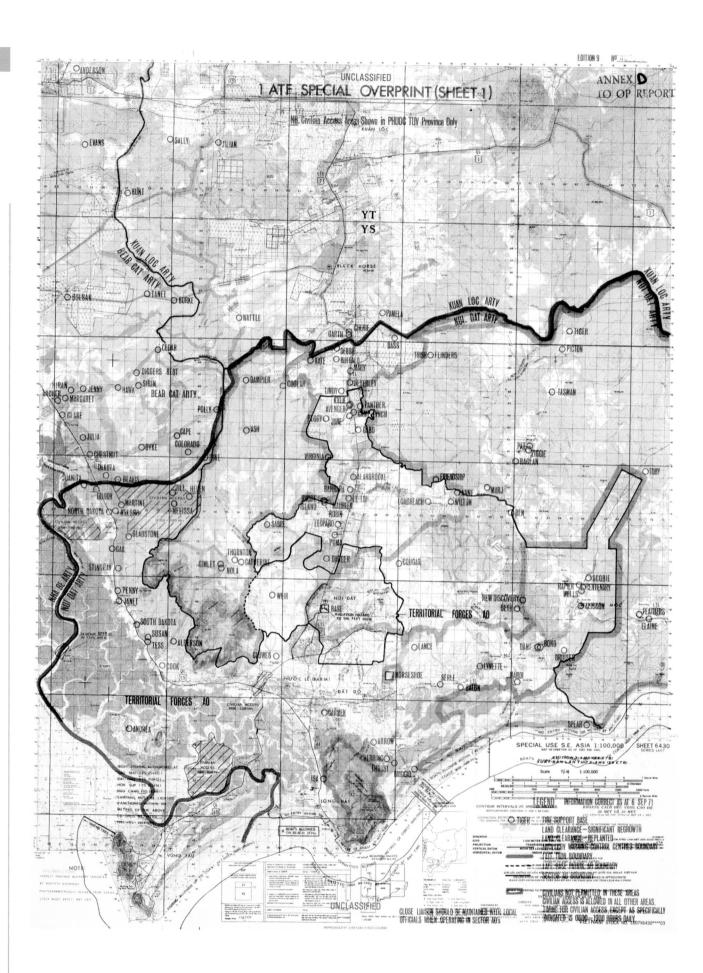

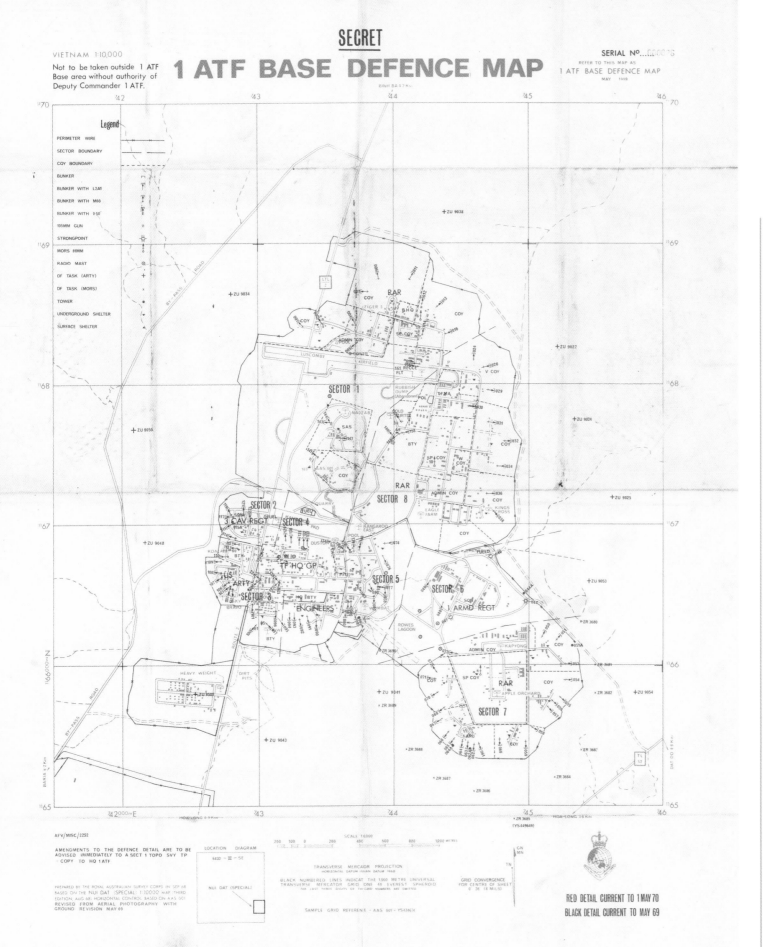

VIETNAM WAR MAP OF THE AUSTRALIAN BASE AT NUI DAT, 1970 The map shows perimeter wire, sector and company boundaries, bunkers, strongpoint, radio masts, towers, underground shelters and surface shelters. Base map in 1968 with red details current to 1 May 1970. Produced by the Royal Australian Survey Corps and endorsed 'Not to be taken outside 1 AFT Base area without authority of Deputy Commander 1 ATF.' LEFT

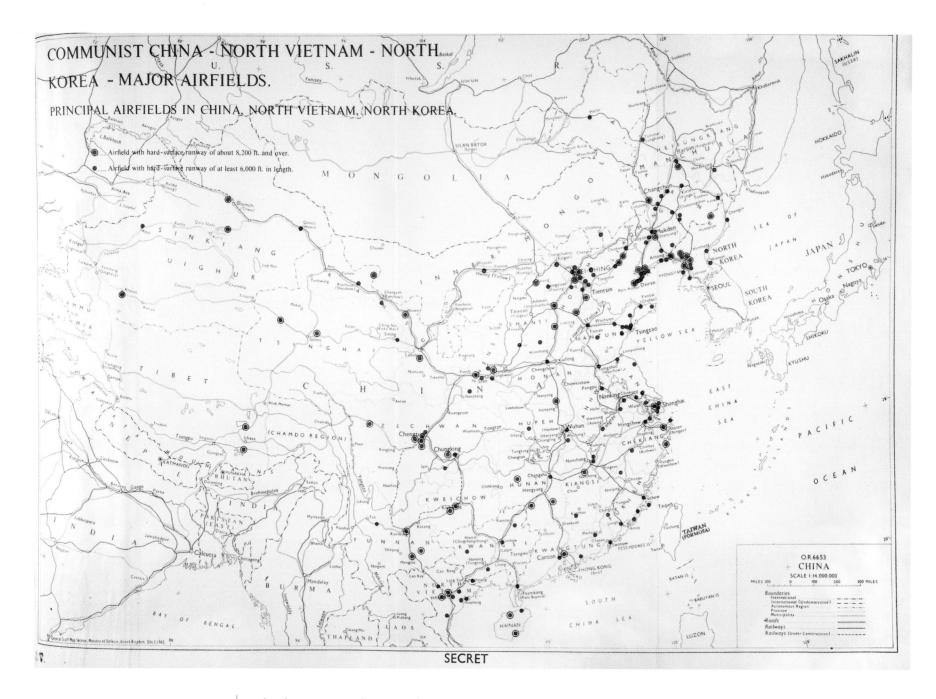

Airfield with hard-surface runway of about 8,200 ft. and over.

Airfield with hard-surface runway of at least 6,000 ft. in length.

O.R. 6653
CHINA
SCALE 1:14,000,000

MILES 100   0   100   200   300 MILES

Boundaries
 International
 International (Undemarcated)
 Autonomous Region
 Province
 Municipality
Roads
Railways
Railways (Under Construction)

SECRET

MAJOR AIRFIELDS IN CHINA, NORTH VIETNAM AND NORTH KOREA, BY THE BRITISH JOINT INTELLIGENCE COMMITTEE, 1968 Concern about the possible broadening out of the Vietnam War was a factor in the strategy of the conflict. The North Vietnamese benefited from air support in opposing American bombers, notably agile Soviet MIG-17s and MIG-21s. However, Soviet surface-to-air missiles, supplied to North Vietnam from 1965, inflicted the greatest damage. The Americans, in turn, benefited from improved equipment and tactics. **ABOVE**

the depiction on the map of Laos' neighbours: Thailand, Cambodia, South Vietnam and Burma. Thus, the domino theory (the theory of Communist advance in stages) was in play and, in this case, was used to support the deployment of 10,000 American marines based in Okinawa.

The Cold War produced particular requirements for mapping, and not only technological ones linked to weaponry. For example, both blocs had major requirements for interoperability. Thus, NATO defined a geographic policy. This introduced the Universal Transverse Mercator (UTM) grid on NATO maps. Subsequently, NATO encouraged digital information exchange.

SURVEYS AND SURVEILLANCE

Mapping for war was greatly affected by the changing character of surveillance and information. Initially, photo-reconnaissance continued the practices of the Second World War. However, the spatial awareness of empire was enhanced by post-war mapping. In the case of Britain, the Colonial Office established the Directorate of Colonial (later Overseas) Survey in 1946 and instructed it to map 900,000 square miles of Africa within ten years, using aerial photography as well as ground surveys. The impressive expertise developed in photo-reconnaissance during wartime was deployed to map large areas previously surveyed only poorly. This capability was especially valuable for

PENTAGON BRIEFING MAP, 19 FEBRUARY 1991
Borrowing an Israeli method successfully employed against the Syrians in 1982, the Americans used decoys to get the Iraqis to bring their radars to full power, which exposed them to attack. In part, the air campaign drew on the ideas of Colonel John A Warden III who suggested a 'five-ring model' of the modern state, each ring linked to a level of activity and a category of target. By attacking the strategic centre, the command and control system, it was claimed, the regime could be defeated with its system paralysed.

However, as mounted in 1991, the air campaign included a focus on all rings apart from population, including the Iraqi armed forces as a whole. The Americans were confident that one of the rings would decisively collapse. Virtually all communications were rendered unusable by the air attack, but the strategic consequences were limited: The Iraqi government did not fall, and it was necessary for the Coalition to launch a ground offensive. Digital photogrammetry was employed to create terrain models for cruise missile attacks on Iraqi targets.

Large-scale air attacks ensured that Iraqi ground defences were left disorganised, isolated and ineffective before the Coalition ground attack began. The psychological impact was reflected in the 87,000 who surrendered and 150,000 deserters, while fewer than 10,000 were killed. The possibility of mounting a cohesive and coordinated defence had been destroyed. This was more significant than the extensive destruction of equipment: the Iraqi forces lost the ability to move. Indicating the unique ability of air power to provide attack in depth, these results were obtained at the expense of the front line of Iraqi formations, and throughout the theatre. As a result, there were no Iraqi units in the rear that could be readily brought forward to replace or reinforce those at the front. Seventeen bridges across the Rivers Tigris and Euphrates were destroyed. **LEFT**

inaccessible terrain. Aerial photography could achieve precise results far more rapidly than ground surveys and was not dependent on local manpower, but it was still a slow process. Later satellite mapping, in contrast, was comprehensive and provided the basis for precise targeting data.

Conflicts in defence of British imperial interests led to the production of topographic maps by Field Survey Squadrons. This happened in Kenya in response to the Mau Mau uprising; in Malaya, also in the 1950s, during the 'Emergency' as Communist insurgents were defeated; and in Borneo during the Confrontation. Similarly, the French were active in map-making in Algeria in 1956–62 as they resisted

insurrection, and the Portuguese in their African colonies. When the British drove the Argentinian invaders from the Falkland Islands in 1982, they suffered from a lack of adequate mapping, including no digital coverage. All available maps of the islands were rapidly bought up. The same happened in Britain with Afghanistan in 2001.

The range of reconnaissance aircraft increased greatly with the introduction of jet aircraft. The American Boeing RB-47 Stratojet, a version of the high-speed and high-altitude B-47E bomber, first flown in 1953, had a range of about 4,000 miles. The availability of Soviet jet fighters, however, threatened these flights, as did improvements to ground-based

PREPARING FOR ATTACK DURING EXERCISE DESERT

SPRING, 2002 Soldiers of the American 3rd
Infantry Division during Exercise Desert
Spring on 18 December 2002. The
Americans focused on Iraq, a definite and
defiant target with regular armed forces,
rather than on the more intangible
struggle with terrorism, which challenged
Western conventions of war making. The
American campaign was to be widely
praised for its manoeuvrist character and
its ability to seize and sustain the
initiative. **RIGHT**

JOINT SERVICE PLANNING AND BRIEFING FOR

SENIOR BRITISH OFFICERS BEFORE THE

DEPLOYMENT OF BRITISH FORCES TO IRAQ, 2003

Officers gather round the map table in
2003 as part of a Joint Service briefing for
senior British officers before the
deployment of British forces to Iraq.
Britain took a secondary part in the
invasion, but an important one because
of the potential for overstretch created
by the relatively small size of the force
deployed by the Americans. On the
ground, 125,000 American combat troops
were the crucial element, compared with
45,000 troops from Britain. The British
played a key role in the south but were
too heavily committed there to offer
effective assistance farther north. The
Permanent Joint Headquarters for the
conduct of British joint operations had
been established in 1996. **FAR RIGHT**

anti-aircraft fire. As a consequence, the U–2 was
introduced by the Americans, with its first flight in
1955. Able to fly at more than 80,000 feet and to
provide accurate pictures, the U–2s provided much
long-range intelligence for strategic intentions and
target selection.

SATELLITES

Aircraft as a source of information were supplemented
by satellites, projected into orbit by rockets after the
first satellite, the Soviet Sputnik, was launched in 1957.
Orbiting satellites offered the potential for obtaining
material, for the radio dispatch of images and for the
creation of a global telecommunications system. In

1960, the first pictures were received from Discoverer
13, the earliest American military photo
reconnaissance satellites. The overlap between
photography and mapping was increasingly
pronounced. Regular satellite images provided the
opportunity to map change on the ground, including
the construction of missile sites. Satellites were too
high to be shot down, as a U–2 aircraft, en route from
Peshawar in Pakistan to Bodø in Norway, had been at
75,000 feet over Sverdlovsk by a Soviet surface-to-air
missile in 1960. Satellites offered the possibility of
frequent overflights and thus of more data. The first
photographs taken by an American satellite were
returned to Earth in 1960 by means of a capsule

ejected from the spacecraft. That year, the National Reconnaissance Office was created as a secret agency designed to coordinate American spy satellites. It became possible to map the entire world at small scale.

The National Photographic Interpretation Center was created by the Americans in 1961 in order to perform photographic analysis as part of the CIA's Directorate of Science and Technology. Photo-reconnaissance was also crucial to the American Army Map Service, which, as a result, delivered large numbers of maps to American forces in the Vietnam War. The Americans enhanced their reconnaissance

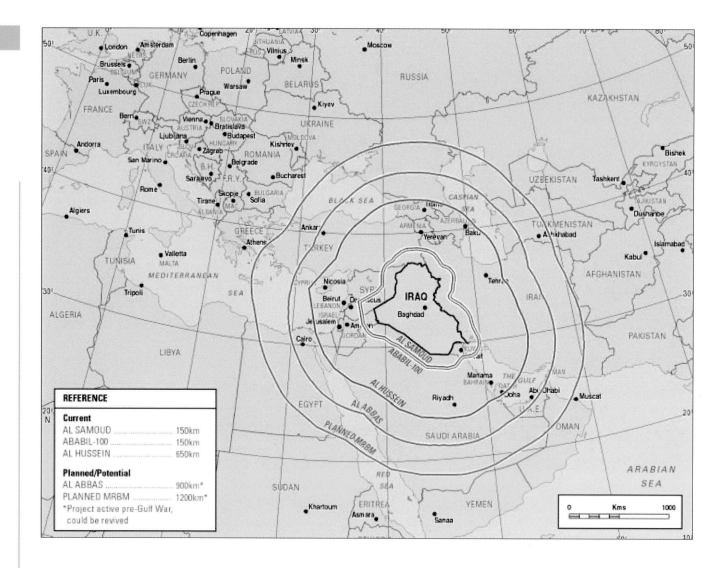

DOSSIER, 2002 This map from the British
dossier of evidence against Iraqi
President Saddam Hussein shows the
alleged range of the country's current
and planned missiles. The dossier claimed
that Iraq was developing missiles capable
of reaching the British Sovereign Base
Areas in Cyprus and NATO members
(Greece and Turkey), as well as Iraq's Gulf
neighbours and Israel. Iraq indeed was
seeking to develop chemical and
bacteriological warheads, but the
reliability and use of Anglo-American
intelligence about Iraq's weapons of mass
destruction were subsequently to be
discredited. **RIGHT**

FACING INSURGENCY IN IRAQ, 2006 An
American infantry officer studies a map
after American and Iraqi commanders
announce at a joint press conference that
their forces will pacify Baghdad district
by district. The large-scale and resilient,
but also complex, insurgency challenged
American control and led to difficult
urban fighting. The Americans became
more effective in counter-insurgency, in
part by rediscovering a relevant doctrine.
A more perceptive response to Iraq's
sectarian divides permitted
differentiating among opponents in
information-driven operations. Initially, in
Iraq, the American forces had not known
whom they were fighting. **FAR RIGHT**

capability with the Lockheed SR-71 Blackbird, which
could fly at Mach 3 outrunning any missile, at over
85,000 feet, and could provide reconnaissance
information on 100,000 square miles in one hour
(North Korea, in comparison, is 46,650 square miles
in area). Moreover, drones brought back high-
resolution photographs from Vietnam. That war saw
the establishment in 1972 of the Defense Mapping
Agency (DMA). While needing massive IT assistance
to access the information acquired, the military has
profited greatly from the use of digital technology for
mapping and for the application of cartographic
information.

The use of such material encouraged secrecy about
cartographic applications. This was an important
aspect of a more general secrecy that was seen, in
particular, with the deliberate distortion of maps, not

least by altering positions, but also by omitting
information. This process was particularly apparent
with the Eastern Bloc during the Cold War, but was
not only seen there. For example, the American
government omitted federal military sites, such as the
North American Aerospace Command, from maps,
and also degraded GPS radio signals for non-military
uses. British maps also disguised facilities. The use of
maps to help targeting was therefore linked to the
more general process of camouflage, with the
distortion of maps a key element of camouflage.

Faced, as a result of aerial surveillance, by a vast
amount of data, the military was also active in
addressing improved systems for data compilation and
usage. The need to analyse vast amounts of
information provided a major spur in the development
of computer systems and then, more significantly, in

level, not least, but not only, by states that lacked this cutting-edge capability. Different but sometimes related to this, the limited mapping of much of the world for much of the century, and notably its first two-thirds, could be a factor. Thus, in 1963–66, the maps used by the British military in their 'Confrontation' (low-level conflict) with Indonesia in Borneo were largely blank for the Indonesian side of the frontier with the former British territories (by 1963 part of Malaysia) that the British were protecting.

Newly-independent states rapidly established their own mapping agencies, in many cases building on previous imperial bodies. For example, the Egyptian Military Directorate was founded in 1954. In most states, military mapping was, and is, highly confidential, for example that of India on its borders with China and Pakistan. Maps in the Indian Army are prepared by the Survey Department which comes under the General Staff. As an aspect of conflict as broadly conceived, maps are also used to assert claims, as with the frontiers of South Asia, and to rename and remake landscapes. For example, Arab villages were taken out of Israeli maps.

MODERN CHALLENGES

The situation was far more limited for insurrectionary groups as they did not have access to the modern facilities for mapping. Conversely, their need for maps was more limited. Moreover, the mapping of insurgency and counterinsurgency struggles, and indeed terrorism and counterterrorism, was problematic, an issue that continues to this day. In this warfare, the notion of control over territory is challenged by forces that cannot be readily described in terms of conventional military units. They seek to operate from within the civilian population and do so not only for cover and sustenance, but also in order to

deny their opponents any unchallenged control over populated areas.

It is extremely difficult to map a situation of shared presence, one in which military or police patrols move unhindered, or suffer occasional sniping and ambushes, and have to consider mines; but, otherwise, control little beyond the ground they stand on. Aerial supply and attack capabilities further complicate the situation. Thus, front lines dissolve and are perceived as dissolved.

These points are particularly relevant as most conflicts today occur as forms of civil war and thus within countries, for example Afghanistan, Iraq and Syria. As a consequence, the situation is complex and is not one readily addressed in terms of cutting-edge capability, whether weaponry or with reference to information availability and cartographic methods. There is no sign that this situation is changing. As modern warfare amply indicates, an understanding of place, while very useful at the strategic, operational and tactical levels of war, is less helpful in enforcing will, the true goal of conflict. Indeed, in some respects, by creating a deceptive sense that other peoples and lands were readily 'knowable', mapping was actively misleading. All sides use maps as a propaganda tool to 'show' their 'liberated' territory. Such maps are an important public relations tool, but can mislead political leaders. For example, the definition of a pacified area in Afghanistan is highly dubious even without the difficulty of assessing the presence of enemy elements.

At the same time, great power confrontations suggest that warfare may take a different form and, as a result, that there may be contrasting cartographic needs. This possibility came to the fore in the mid-2010s, first when greater tension between China and Japan led to the possibility of a confrontation between their improvement. Computer systems, in turn, proved particularly significant in new cartographic techniques. The United States was much more adept than the Soviet Union in developing computer capacity, not least in image correlation and aerial triangulation so as to produce digital terrain methods.

Satellite information also came to serve as the basis for enhanced weaponry. The US Department of Defense developed a global positioning system (GPS) that depended on satellites, the first of which was launched in 1978. Automatic aiming and firing techniques rested on accurate surveying. 'Smart' weaponry, such as guided bombs and missiles, make use of precise mapping in order to follow predetermined courses to targets actualised for the weapon as a grid reference. Cruise missiles use digital terrain models of the intended flight path. The American BGM-109 Tomahawk Block III cruise missile, with its improved accuracy and stand-alone GPS guidance capability, was first used in the September 1995 Bosnia operation, achieving a success rate of more than 90 per cent. The Soviet Union sought to match the American GPS system with its own Global Navigation Satellite System, which became operational in 1995 and, in 1996, reached its full design specification of 24 satellites.

In the 1991 First Gulf War, the American DMA alone produced more than 110 million maps for its forces and, benefiting from digital photography, the Americans used cruise missiles with great accuracy and effect against Iraq. Moreover, precise positioning devices interacted with American satellites to offer an effective GPS that was employed with success by Allied tanks in their combat with Iraqi counterparts. In addition, satellite information helped in the rapid production of photo-maps. Geographical Information System software provides instantaneous two- and

three-dimensional views of battlefields. The devastating firepower of modern weapons, and their capacity to recreate the landscape, encouraged an emphasis on precision, especially in an urban environment.

LOCAL LEVEL
Alongside high-specification mapping came the reality of more ordinary maps used for struggles at the local

IRAQI OPERATION AGAINST ISIS, 2015 Iraqi Defence Minister Khaled al-Obaidi provides the details of an operation on the map as he visits military units in the Taci District of Saladin on 26 February 2015. The Iraqi army was preparing for an operation to be carried out against the forces of ISIS (Islamic State of Iraq and Syria). The dislike and mistreatment of the Shi'a-dominated government for the Sunni minority and the resentment generated among the latter had ensured that the army was unable to resist ISIS when it advanced on Mosul in June 2014. ISIS there seized money and equipment, and its transition from an insurgency to a de facto state gathered pace.

ABOVE

DIGITAL MAP USE IN IRAQ, 2015 A commander from the pro-government Badr Brigade militia looks over a map on a tablet while touring the front line with ISIS fighters on 11 April 2015 in Ebrahim Ben Ali, in Anbar Province, Iraq. Shia militia and Iraqi government troops were preparing for an assault on ISIS forces in Anbar, much of which was captured by ISIS forces the previous year. RIGHT

AYN AL-ARAB, SYRIA, 2014 Staffan de Mistura, United Nations special envoy for Syria, shows a map of the town of Ayn al-Arab, known as Kobane by the Kurds, during a press conference at the UN office in Geneva on 10 October 2014. He urged Turkey to allow Kurds to cross into Syria to protect Kobane against ISIS fighters, who had seized part of the key Syrian border town despite American-led air strikes that led to heavy casualties among the fighters. Tension between the Turkish government and the Kurds greatly complicated the issue. FAR RIGHT

China and Japan's ally, the United States, and, secondly, when the crisis between Russia and Ukraine in 2014 escalated to include the possibility of Russian expansionism at the expense of Estonia, Latvia and/or Lithuania, with the danger that such activity might lead to conflict with NATO. These possibilities led to an emphasis on the mapping of conventional confrontation and conflict, but such mapping was affected by the new possibilities created by developing technology, notably with cyber warfare. In the context of real-time mapping in response to the closure of the gaps between surveillance, decision and firing system, the prospect of such mapping being affected by attacks on communication systems, whether satellites or

computers, was one that threatened to plunge opponents into a cartographic void. Thus, ironically, the very enhanced capability that appeared to stem from cartographic improvement and application also threatened a vulnerability that was far greater in type than that posed by attacks on air reconnaissance assets in the two world wars and the Cold War.

This situation was certainly far from static. It was unclear how far drones, and notably micro-drones, would offer an enhanced reconnaissance capability, and also how effective cyber warfare might be. Nevertheless, the dynamic, but also vulnerable, character of military mapping, and its usage, were both clearly in play in the 2010s.

INDEX

PICTURE CREDITS

The author and publishers are grateful to the following for permission to reproduce images. While every effort has been made to contact the copyright holders, if any credits have been omitted in error, please do not hesitate to contact us.

2 © Arterra Picture Library / Alamy Stock Photo
6 Print Collector/Getty Images
8-9 Heritage Images/Getty Images
10 Zev Radovan/www.BibleLandPictures.com/Alamy Stock Photo
11 Corbis
12 De Agostini Picture Library/M. Seemuller/Bridgeman Images
13 Hulton Archive/Getty Images
14-15 © Reproduced with the kind permission of the Holkham Innys Collection
16-17 Imagno/Getty Images
19 © RMN-Grand Palais (domaine de Chantilly)/René-Gabriel Ojéda
20 Bibliotheque Mazarine, Paris, France/Bridgeman Images
21 Bibliotheque Mazarine, Paris, France/Archives Charmet/Bridgeman Images
22 Images@Alteamaps.com
24 Photo Scala, Florence
26 © The British Library Board
27 Bibliotheque Nationale, Paris, France/Archives Charmet/Bridgeman Images
28-29 © RMN-Grand Palais (Château de Pau)/René-Gabriel Ojéda
30 Images@Alteamaps.com
31 Korean Army Museum
32-33 DEA / G.Nimatallah/Getty Images
34-35 Bibliotheque Nationale, Paris, France/Flammarion/Bridgeman Images
36 Private Collection/Bridgeman Images
37 © The British Library Board
38 N/A
39 Heritage Image Partnership Ltd./Alamy Stock Photo
41 Alex Ramsay/Alamy Stock Photo
42-43 Images@Alteamaps.com
44-45 Hessisches Staatsarchiv Marburg

46 Bibliothèque nationale de France
48 Private Collection/Bridgeman Images
49 Imagno/Getty Images
50 akg-images
51 Bibliotheque Nationale, Paris, France/Archives Charmet/Bridgeman Images
52-53 © The British Library Board
54 Museo Correr, Venice, Italy/Bridgeman Images
55 MS Russ 72 6, Houghton Library, Harvard University.
56-57 DEA / A. Dagli Orti/Getty Images
59 INTERFOTO/Alamy Stock Photo
60 Images@Alteamaps.com
62 Images@Alteamaps.com
63 Reproduced by permission of the National Library of Scotland
64-65 Images@Alteamaps.com
66 Images@Alteamaps.com
68 Private Collection/Bridgeman Images
69 © British Library Board. All Rights Reserved/Bridgeman Images
70-71 Library of Congress
72 Courtesy of the National Library of Ireland
75 Bibliotheque Nationale de Cartes et Plans, Paris, France/Archives Charmet/Bridgeman Images
76 Kyujanggak Institute For Korean Studies
77 Library of Congress
78 © The British Library Board
80 akg-images
82-83 akg-images
84 Staatsbibliothek zu Berlin/bpk
85 Staatsbibliothek zu Berlin/bpk
86 akg-images
88 Images@Alteamaps.com
89 © The British Library Board
90 British Library/akg-images
91 Library of Congress
92 Library of Congress
93 Library of Congress
94 Time Life Pictures/Mansell/The LIFE Picture Collection/Getty Images
95 Library of Congress
96 Library of Congress
96 Library of Congress
97 © The British Library Board
98 © The British Library Board

99 © The British Library Board
100-101 Universal History Archive/Getty Images
103 © RMN-Grand Palais (musée des châteaux de Malmaison et de Bois-Préau)/Gérard Blot
104 Library of Congress
105 © The British Library Board
106 National Army Museum, London/Bridgeman Images
108 © The British Library Board
109 Fine Art Images/Heritage Images/Getty Images
110 Images@Alteamaps.com
112 MAP RM 343, http://nla.gov.au/nla.obj-231230934, National Library of Australia
113 Heritage Image Partnership Ltd./Alamy Stock Photo
114 © The British Library Board
116 MAP RM 364, http://nla.gov.au/nla.obj-231238361, National Library of Australia
117 MAP RM 366, http://nla.gov.au/nla.obj-231238767, National Library of Australia
118 DEA Picture Library/Getty Images
119 MAP RM 375, http://nla.gov.au/nla.obj-231242233, National Library of Australia
120 Pictures From History/akg-images
121 Library of Congress
122 Library of Congress
123 Library of Congress
124 Library of Congress
126 Library of Congress
128 Library of Congress
129 MAP NK 6574 A, http://nla.gov.au/nla.obj-230867595, National Library of Australia
130 MAP Ra 148 [Plate 11], http://nla.gov.au/nla.obj-230868270, National Library of Australia
131 akg-images
132 MAP RM 359/2, http://nla.gov.au/nla.obj-231236931, National Library of Australia
133 © The British Library Board
134 Library of Congress
135 Library of Congress
135 The National Archives
136 Science & Society Picture Library/Getty Images
137 MAP RM 2349/6, http://nla.gov.au/nla.obj-231801893, National Library of Australia

138 Bibliotheque Nationale, Paris, France/Bridgeman Images
139 The National Archives
140 Library of Congress
141 Album/Documenta/akg-images
142 Niday Picture Library/Alamy Stock Photo
143 © The British Library Board
144 Library of Congress
145 The National Archives
146 © The British Library Board
147 Library of Congress
148 Art Media/Print Collector/Getty Images
149 Library of Congress
150-151 Ullstein Bild/Getty Images
153 MAP G7831.S6 [1905], http://nla.gov.au/nla.obj-92730161, National Library of Australia
154 Public domain
155 MAP G6851.S65 [1913?], http://nla.gov.au/nla.obj-234558179, National Library of Australia
157 Private Collection/Bridgeman Images
158 © Illustrated London News Ltd./Mary Evans
160 Library of Congress
161 bpk
162 The National Archives
163 © The British Library Board
164 MAP RM 4782, http://nla.gov.au/nla.obj-232524042, National Library of Australia
165 Chronicle/Alamy Stock Photo
166 John Frost Newspapers/Alamy Stock Photo
167 Photo12/UIG
168 Popperfoto/Getty Images
169 MAP G7501.S65, http://nla.gov.au/nla.obj-234266143, National Library of Australia
170 Fototeca Gilardi/Getty Images
171 Library of Congress
172 Library of Congress
173 Library of Congress
174-175 Universal History Archive/Getty Images
177 MAP G7824.S5A4 1932, http://nla.gov.au/nla.obj-233795656, National Library of Australia
178 MAP G7801.S7 1937, http://nla.gov.au/nla.obj-234701228, National Library of Australia
179 MAP G7821.S7 1938, http://nla.gov.au/nla.obj-240156402, National Library of Australia

180 The National Archives
181 Album/Oronoz/akg-images
182 Past Pix/Getty Images
183 © IWM (COL 238)
184 © The British Library Board
185 © The British Library Board
186 akg-images
187 MAP G3201.S7 1942, http://nla.gov.au/nla.obj-234716550, National Library of Australia
188 © Illustrated London News Ltd./Mary Evans
189 akg-images
190 akg-images
191 Roger Viollet/Getty Images
192 The National Archives/SSPL/Getty Images
193 The National Archives/SSPL/Getty Images
194 Album/Prisma/akg-images
195 Library of Congress
196 MAP G5701.S7 1944, http://nla.gov.au/nla.obj-234713882, National Library of Australia
197 Library of Congress
198 MAP G7964.T6 1946, http://nla.gov.au/nla.obj-233357691, National Library of Australia
199 The National Archives
200-201 Gilles Bassignac/Getty Images
203 Library of Congress
204 Sputnik/akg-images
205 The National Archives
206 MAP G8021.R1 1971, http://nla.gov.au/nla.obj-236169716, National Library of Australia
207 MAP G8021.R1 1970, http://nla.gov.au/nla.obj-242208596, National Library of Australia
208 The National Archives
209 Mark Reinstein/Corbis
210 Ron Haviv/VII/Corbis
211 © Crown copyright. IWM (OP-TELIC 03-010-01-052)
212 PA Archive/PA
213 Ahmad Al-Rubaye-Poo/Getty Images
215 Ali Mohammed/Anadolu Agency/Getty Images
216 John Moore/Getty Images
217 Fabrice Coffrini/AFP/Getty Images
Front endpaper Hulton Archive/Getty Images
Back endpaper Library of Congress

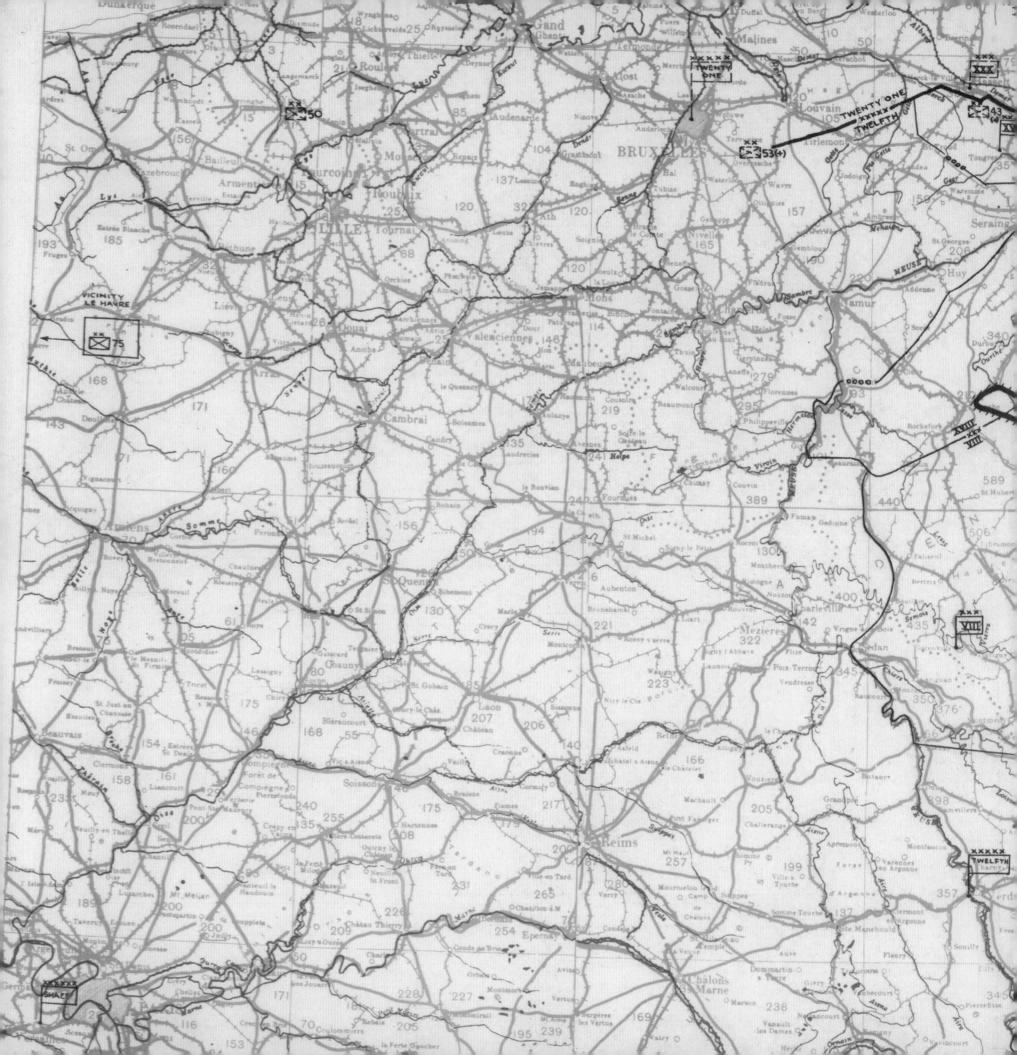